Introduction to Chemical Principles

A Laboratory Approach

Fifth Edition

Introduction to Chemical Principles

A Laboratory Approach

Fifth Edition

Susan A. Weiner

Department of Chemistry
West Valley College
Saratoga, California

Edward I. Peters

Formerly of
Department of Chemistry
West Valley College
Saratoga, California

Saunders College Publishing
Harcourt Brace Jovanovich College Publishers

Ft. Worth Philadelphia San Diego New York Orlando Austin San Antonio Toronto Montreal London Sydney Tokyo

WARNINGS ABOUT SAFETY PRECAUTIONS

Some of the experiments contained in this Laboratory Manual involve a degree of risk on the part of the instructor and student. Although performing the experiments is generally safe for the college laboratory, unanticipated and potentially dangerous reactions are possible for a number of reasons, such as improper measurement or handling of chemicals, improper use of laboratory equipment, failure to follow laboratory safety procedures, and other causes. Neither the Publisher, nor the Authors can accept any responsibility for personal injury or property damage resulting from the use of this publication.

Printed in the United States of America.

Weiner & Peters: _Introduction toChemical Principles: A Laboratory Approach_, fifth edition.

0-03-024828-0

567 023 7654321

Preface

This laboratory manual is designed for a one-semester or one-quarter introductory chemistry course or for a general chemistry course for nonscience majors. There are more experiments than can be completed in a single term. This gives the instructor a wide variety from which to assemble a laboratory program best suited for the particular course. Earlier editions of the manual have been used with different textbooks, including *Introduction to Chemical Principles* and *Basic Chemical Principles,* by Edward I. Peters.

Experiments 12 and 13 are new in this edition. These two, and several others, are microscale (small scale) experiments. This method increases safety, reduces the cost and disposal of chemicals and allows for shorter completion times. Microscale experiments are marked with the symbol Ⓜ in the Table of Contents for easy identification. Experiment 15 from the 4th Edition is omitted, due to the increased concern about exposure to mercury.

As in the 4th Edition, the data and report sheets are printed in duplicate, one identified as a work page and the other as a report sheet. The students are directed to enter data into the work page during the experiment, then copy the finished data and calculations into the report sheet, resulting in a clean and neat report.

Report sheets are long enough to give adequate presentation of observations and results, but short enough to be graded easily and rapidly. Each laboratory experiment is independent of all others, except Experiments 23 and 24. All experiments can be completed in a three-hour laboratory period, including prelaboratory discussion. The Instructor's Manual includes lists of necessary chemicals and equipment, data to be expected, answers to Advance Study Assignments, and miscellaneous suggestions.

A student finishing a laboratory program based on this book will have become familiar with many laboratory operations and will have learned how to collect and analyze experimental data. These skills will be a strong foundation for further coursework in general chemistry or other college-level science curriculum.

Susan A. Weiner

Edward I. Peters

Contents

Ⓜ indicates microscale (small scale) experiments

Safety in the Laboratory

A chemistry laboratory can be, and should be, a safe place to work. Accidents can be prevented if you think about what you are doing at all times, use good judgment, observe safety rules, and follow directions. In addition to the rules below, comments appear in each experiment to alert you to probable hazards, including specific instructions on how to protect yourself and others against injury. Be sure to read these and keep the warnings in mind as you perform each experiment. Do not deviate from the procedures given in this book unless you are instructed to do so.

THERE IS NO SUBSTITUTE FOR SAFETY IN THE LABORATORY. Learn and observe these safety rules at all times:

1. Eye protection (goggles, safety glasses) must be worn by all students when working in the laboratory. This includes clean-up times, and times when you yourself may not be working on an experiment, but someone near you is.
2. Do not eat, drink, or smoke in the laboratory.
3. Do not taste any chemical.
4. Purses, sweaters, lunch bags, backpacks, and extra books should be stored in designated areas, but not in the laboratory working area. Backpacks, in particular, should not be on the floor near your laboratory desk.
5. Shoes must be worn in the laboratory at all times. Bare feet are prohibited.
6. Long hair should be tied back or pinned up, so it will not fall into chemicals or flames.
7. Do not work in the laboratory alone. An instructor or teaching assistant must be present.
8. Never perform any unauthorized experiment.
9. Before leaving the laboratory, wipe the desk top and wash your hands with soap and water.
10. If an accident occurs in the laboratory, no matter how minor, report it to the instructor immediately.
11. All experiments or operations producing or using chemicals that release poisonous, harmful, or objectionable fumes or vapors MUST be performed in the fume hood.
12. Never point the open end of a test tube at yourself or at another person.
13. If you want to smell a substance, do not hold it directly to your nose; instead, hold the container a few centimeters away and use your hand to fan the vapors toward you.
14. Hot glassware and cold glassware look alike. If you heat glass and put it down to cool, do not pick it up too soon. Do not put hot glassware where another person is apt to pick it up.
15. When inserting a glass tube, rod, or thermometer into a rubber tube or stopper, protect your hands by holding the material with gloves or layers of paper towel. Lubricating the glass with water or glycerine is helpful.
16. When diluting acids, always add the acid to water, never water to the acid.
17. Most organic solvents are flammable. Keep these liquids away from open flames.
18. Do not pour organic solvents down a sink in the open laboratory. Dispose of them as directed by your instructor, or down a drain in the fume hood. Flush with plenty of water.
19. When disposing of liquid chemicals or solutions in the sink, flush with large quantities of water.

20. Do not dispose of matches, paper, or solid chemicals in the sink. Matches, after you are sure they are extinguished, and paper should be discarded into a wastebasket. Solid chemicals should be disposed of in whatever facility is provided in your laboratory.

21. Do not put broken glassware into wastebaskets. Dispose of it in designated places.

22. If you should have skin contact with any harmful chemical, flush the contact area with large quantities of water. Have a nearby student call the instructor for aid.

23. If you spill any chemical, solid or liquid, be sure to clean it up so another student does not come into contact with it and perhaps be injured by it.

24. Chemical characteristics, hazard levels, and safety instructions for the chemicals you use in the laboratory are described in Material Safety Data Sheets (MSDS) that are generally available in the laboratory. Follow directions given by your instructor in regard to these sheets. Pay close attention to particular safety precautions your instructor talks about before you begin each experiment.

PREVENTING CONTAMINATION OF CHEMICALS

In order to conduct experiments successfully, you must avoid contaminating the chemical reagents you use, or reagents that will be used by other students after you. The following procedures will help minimize the possibility of contamination:

1. After washing glassware, always use a final rinse of deionized or distilled water.
2. Avoid handling more than one reagent bottle at a time, so you do not interchange their stoppers by mistake.
3. When selecting a reagent bottle, read the label twice to be sure you have the chemical you want.
4. Do not lay tops of reagent bottles or stoppers on the laboratory bench.
5. Use separate spatulas to remove different solid chemicals from their bottles.
6. Never remove a liquid reagent from a stock bottle with an eye dropper. Pour a small portion into a clean, dry beaker, and use your eye dropper to remove the liquid from the beaker.
7. When a quantity of a chemical is removed from its original container, whether it is a solid or a liquid, *do not* return any excess to the stock bottle. Dispose of the unused portion as directed by your instructor.
8. Never weigh a chemical directly on a balance pan. Use a preweighed container. Weighing paper is acceptable for most solid chemicals.
9. Some chemicals react with some stoppers. If you are going to store a chemical or solution in a bottle other than its original container, be sure the stopper you select (glass, rubber, cork) is suitable for that substance.
10. Never leave a stock bottle uncovered. Be sure you cover the bottle with the proper cover.

Common Laboratory Equipment

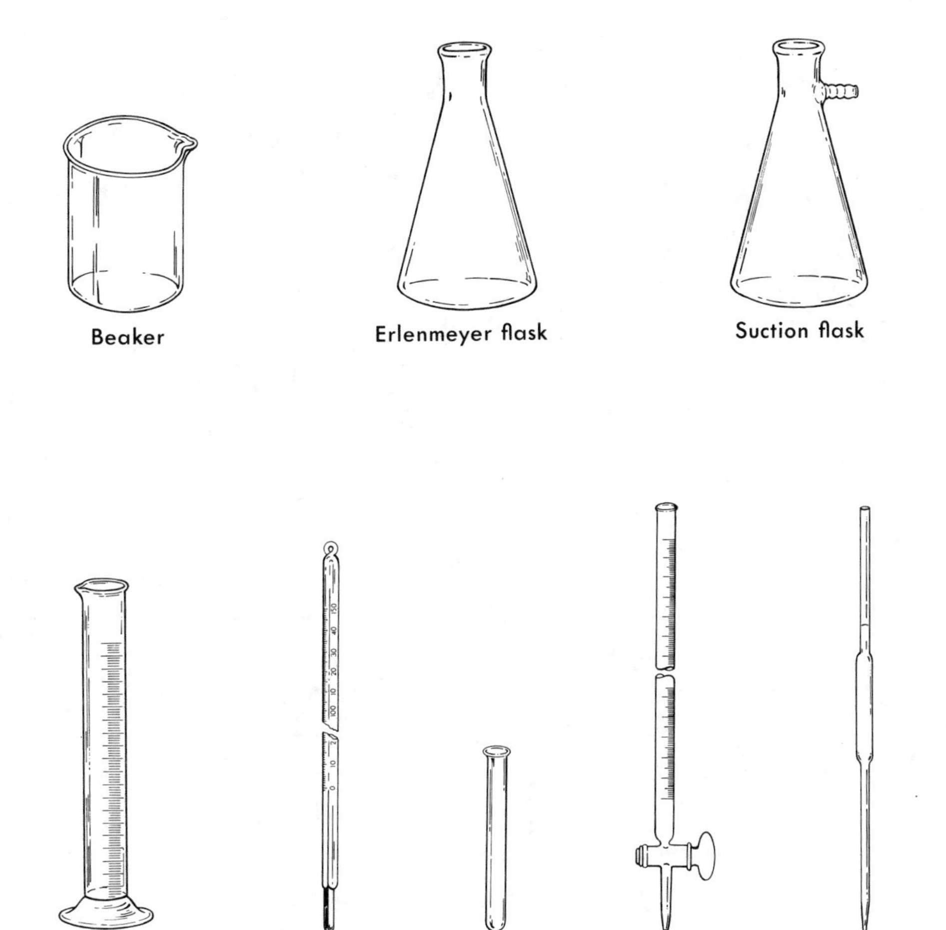

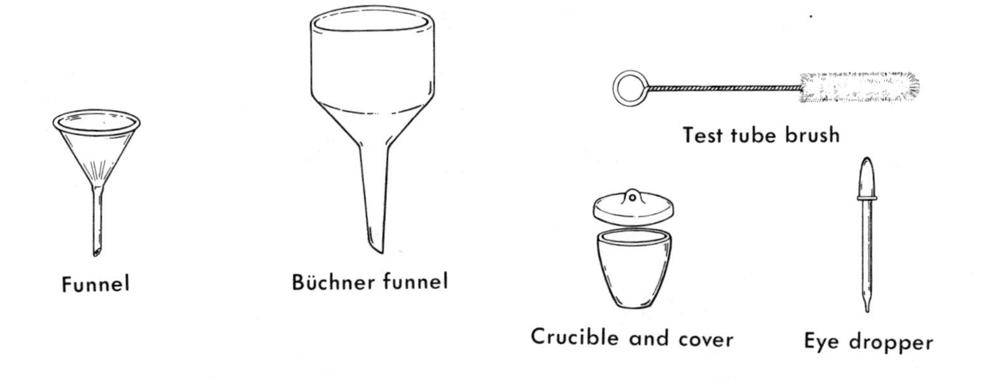

Common Laboratory Equipment

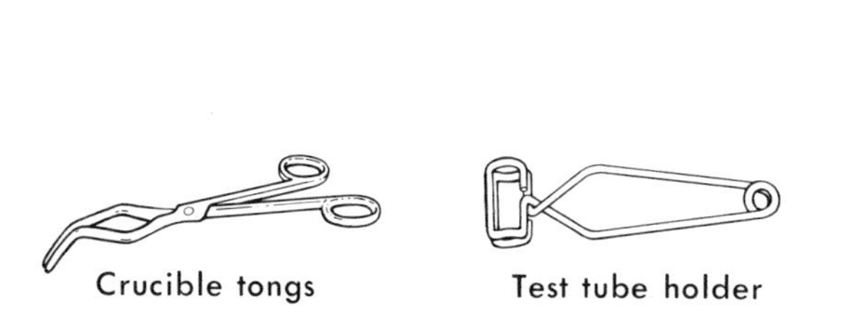

Crucible tongs

Test tube holder

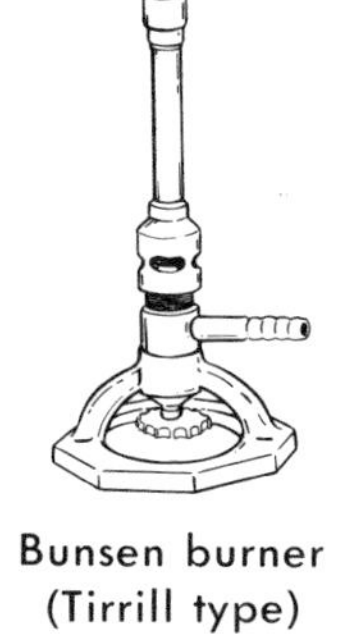

Bunsen burner
(Tirrill type)

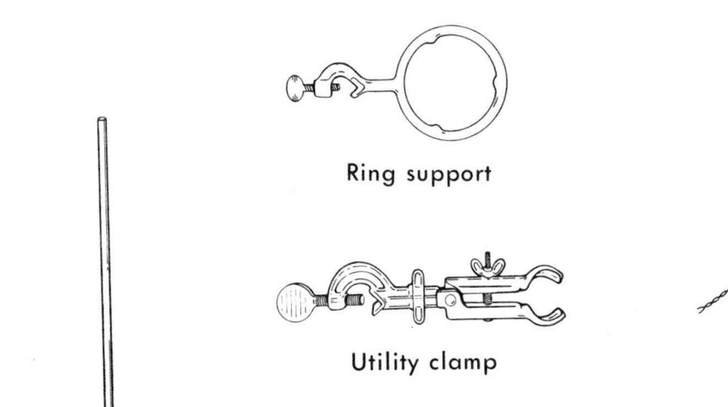

Ring support

Utility clamp

Ring stand with support

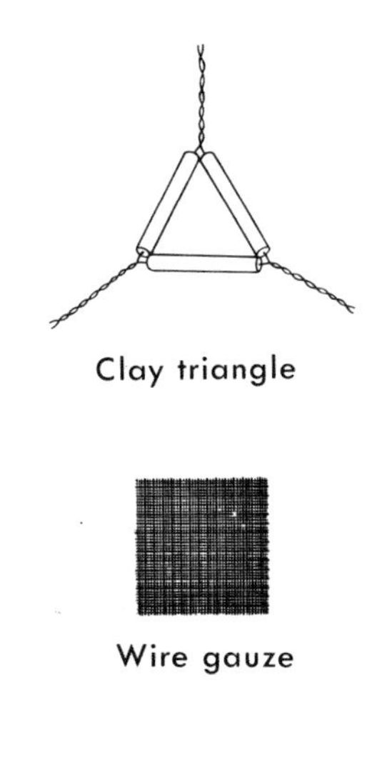

Clay triangle

Wire gauze

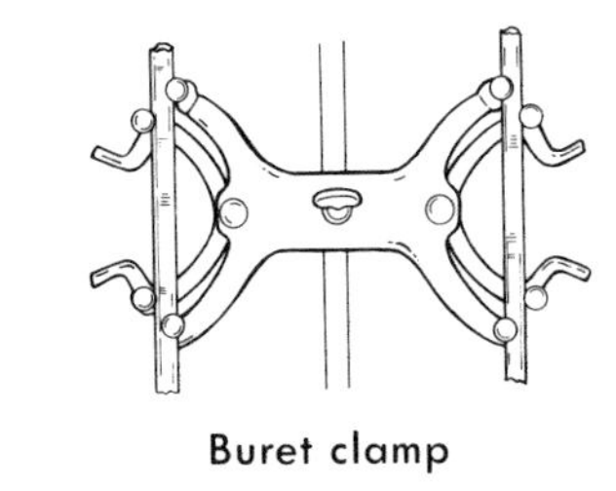

Buret clamp

Evaporating dish

Watch glass

Tripod

Laboratory Procedures

The techniques found in many laboratory operations are so common that your instructions say simply, "Do this . . ." with the assumption that you know exactly how to do it. For beginning students this assumption is often wrong. This may be your first opportunity to conduct a routine operation, and you may have questions about how to do it. This section discusses some of these methods.

HANDLING SOLID CHEMICALS

Your first step in taking a solid chemical is to read the label very carefully to be sure that you get the chemical you want. The names and formulas of different chemicals may be almost identical. For example, sodium sulfate is Na_2SO_4, and sodium sulfite is Na_2SO_3. The names differ by one letter, and the formulas differ by 1 in a subscript. We strongly recommend that you read all chemical names and formulas twice in the laboratory manual and twice again on the supply bottle.

When you need a chemical, take the container in which you will place it to the station from which the chemical is distributed. Transfer the chemical to the container there. Do not take a supply bottle to your work area.

Solid chemicals are generally distributed in wide-mouth, screw-cap bottles. If the substance is "caked" and doesn't flow easily, screw the cap on tightly and strike the bottle sharply against the palm of your hand. If this doesn't loosen the chemical, remove the cap and scrape the packed solid with the scoop you will use to remove the substance from the bottle. Having loosened the solid somewhat, you can often get it to flow freely by recapping the bottle and hitting it against your hand.

When you remove a cap from a bottle, place it on the desk with the top, or outside, of the cap down. This prevents contaminating the inside of the bottle from a dirty desk when the cap is returned to the bottle. Using a clean scoop, remove the amount of chemical you need. If you are transferring a closely controlled quantity of chemical, you can regulate the flow of solid from your scoop by holding the scoop over the receiving container and tapping your hand gently, as in Figure LP–1. If you have removed too much chemical, *do not return the excess to the bottle, rather throw it away.* The waste from this procedure is less of a problem than the contamination that will eventually occur if excess chemicals are returned to supply bottles. It follows that you should judge your requirement carefully and take no more chemical than you need.

After you have transferred the chemical you need, return the cap to the bottle and tighten it securely. If any solid has been spilled on the desk, clean it up before leaving the distribution station.

HANDLING LIQUID CHEMICALS

The general procedures for handling solid chemicals apply to liquids, too. Specifically, (1) double-check the name and/or formula of the chemical you require and the chemical you get; (2) take your container to the distribution station, rather than taking the supply bottle to your work area; (3) do not place the cap or stopper of a supply bottle on the desk in such a way that the inside of the cap touches the desk; (4) if you remove too much liquid from the supply bottle, do not return it, but throw it away; (5) be sure to return the

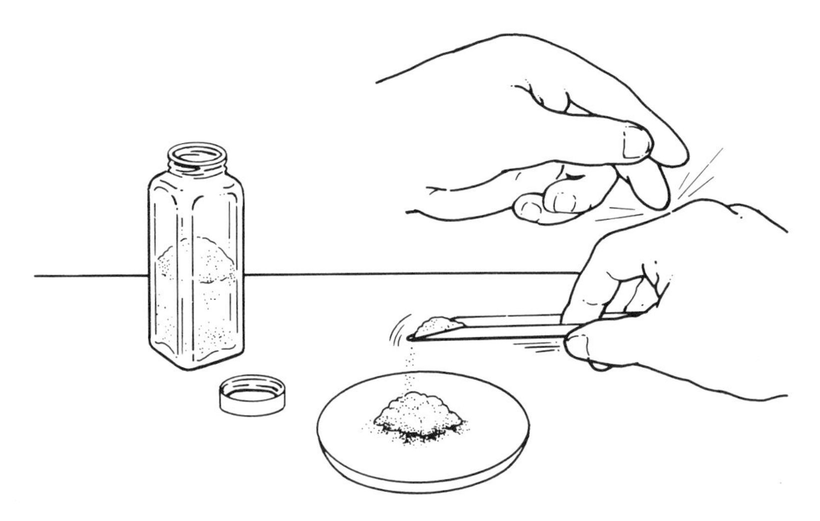

LP–1. Transferring solid chemicals.

cap or stopper to the supply bottle when you are finished; and (6) wipe up any liquid that may have spilled. In regard to (3) above, liquids are frequently distributed in bottles having caps with vertical surfaces that allow you to hold the cap between the fingers of the same hand that is holding the bottle while pouring, as seen in Figure LP–2.

Figure LP–2 also shows the technique for controlling the flow of liquid from a bottle by pouring down a stirring rod. This technique may also be used when pouring from a beaker, as shown in Figure LP–3.

Liquids are frequently distributed from bottles fitted with eye droppers. When using a dropper for removing the liquid, be sure to hold the dropper vertically *with the rubber bulb at the top* so that the liquid does not drain into the bulb and become contaminated. If you are required to take a small quantity of liquid from a bottle not fitted with a dropper and wish to use your own, do not place your dropper into the supply bottle. The proper procedure is to pour some of the liquid into a small beaker and then use your dropper to transfer the liquid from the beaker to your container. Excess liquid should be thrown away, as noted above. Estimate your needs carefully so the excess can be kept to a minimum.

Many liquids used in the laboratory are flammable, many release harmful vapors, and many have both of these dangerous properties. When working with such chemicals, it is best to work in a fume hood. When disposing of such chemicals, always follow the specific procedures established in your laboratory. If they are to be poured down the drain, use a drain in a fume hood, not a sink drain in the open laboratory. If a liquid is flammable, do not use it anywhere near an open flame. Vapors from your liquid could drift to the flame and become an invisible wick by which the flame could travel right back to your liquid and cause a fire.

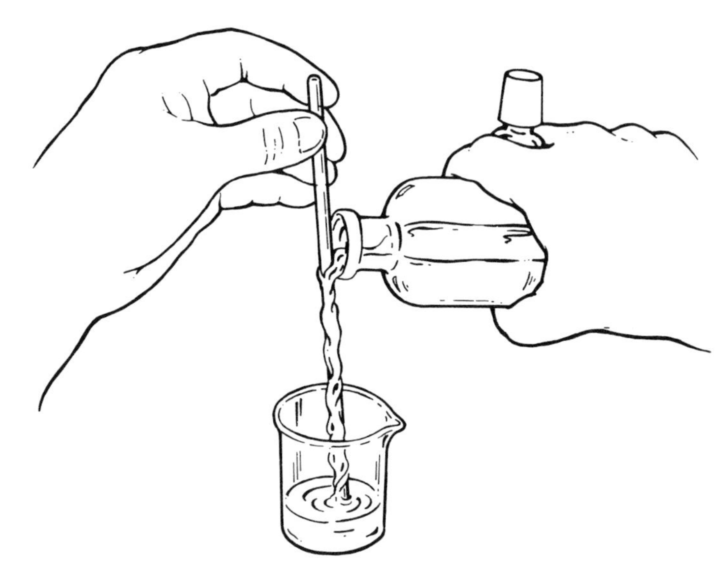

LP–2. Pouring liquids from a bottle.

LP–3. Pouring liquids from a beaker.

QUANTITIES OF CHEMICALS

Most chemical quantities identified in this book are approximate quantities that are practical for the sizes of beakers, test tubes, and other containers you will use. If the quantity you take falls within 10 percent of the amount called for, it will be satisfactory. It is therefore unnecessary for you to try to measure out "exactly" the amount specified. In fact, trying to get that exact amount is a waste of time, both your time and the time of other students who will be delayed because you tie up a balance for so long.

While using an exact quantity of a chemical is not important, knowing as accurately as possible the quantity actually used is essential if that quantity becomes a part of your calculations. You will recognize this requirement if your instructions call for so many milliliters of a liquid "estimated to the nearest 0.1 mL," or to "measure 1.5 grams of a solid on a milligram balance." The first tells you to pour into a graduated cylinder a quantity of liquid that is within about 10 percent of the amount specified, and then to measure and record that quantity to the nearest 0.1 mL. The second instruction may be interpreted as, "Take between 1.35 and 1.65 grams of a chemical and then measure and record the quantity taken to the nearest milligram."

Several experiments in this book require "about 1 to 2 mL" of a liquid, usually to be placed in a test tube. Again the exact quantity is not important, and it is a waste of time to measure it with a graduated cylinder. Most eye droppers deliver drops of such size that there are about 20 drops to the milliliter; and the total volume drawn into a dropper by one squeeze of the bulb is about 1/2 milliliter. One milliliter therefore can be estimated simply as two droppersfull.

READING VOLUMETRIC GLASSWARE

When a liquid is placed into a glass container it forms a **meniscus,** a curved surface that is lower in the middle than at the edge. Volumetric laboratory equipment is calibrated to measure volume by sighting to the *bottom* of the meniscus, as shown in Figure LP–4. Notice that it is essential that the line of sight be perpendicular to the calibrated vessel if you are to read it accurately. It is also important that you hold the vessel vertically.

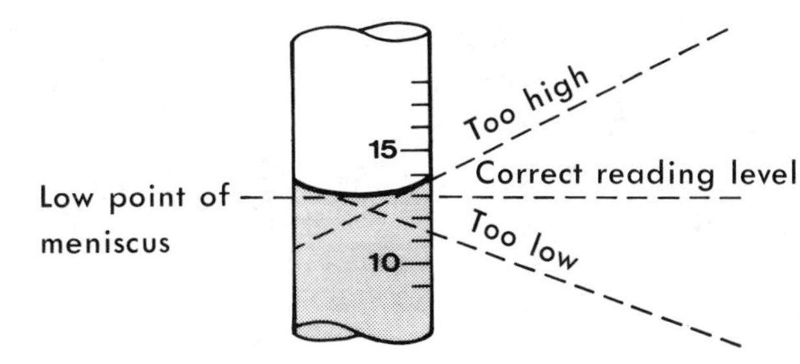

LP–4. Reading the volume of a liquid.

Four types of calibrated glassware are used in the experiments in this book. The most accurately calibrated are volumetric pipets and burets used in Experiments 23 and 24. Most of your volume measurements will be made in graduated cylinders. Their main purpose is to measure volumes and they are designed and calibrated accordingly. Beakers and Erlenmeyer flasks made by some manufacturers are also "calibrated," even though the function of these items has nothing to do with measuring volume. The calibrations on beakers and flasks give only *very rough* indications of volume up to a certain level in the vessel. Volumes estimated by these calibrations should *never* be used in calculations.

MEASURING MASS—WEIGHING

You will no doubt receive specific instructions on the use of chemical balances in your laboratory. No attempt will be made to duplicate those instructions here. Instead, comments will be limited to some general suggestions, plus identification of a term that has special meaning throughout this book.

Chemicals are *never* weighed directly on the pan of a laboratory balance. Instead, the mass is determined by a process known as **weighing by difference**. A suitable container—a small beaker, or perhaps a test tube that is to be used in the experiment—is weighed empty on the balance. The desired chemical is added to the container, and the total mass of the combination is determined. By subtracting the mass of the empty container from the mass of the container plus chemical, you find the mass of the chemical.

Throughout this book the word **container** is used to include any and all objects that pass through the entire experiment unchanged in mass. In addition to a test tube, for example, you might include in the mass of the "container" a test-tube holder by which the test tube is suspended on a balance during weighing, or the mass of a beaker in which the test tube is held for weighing. In one experiment the mass of a liquid is measured in a graduated cylinder that is covered with a piece of plastic film. The film is weighed with the empty cylinder, and their combined masses make up the mass of the "container." In the various experiments where you see the "container" identified, the word has the meaning given in this paragraph.

Sometimes students use containers that are not actually part of the experiment in taking samples of solid chemicals. Most common is the practice of placing a piece of paper on the pan of a balance, transferring the required quantity of chemical to the paper, and then transferring it to the vessel to be used in the experiment. If you use this technique to obtain a measured mass of the chemical, your first weighing should be of the paper with the chemical on it. Then transfer the chemical, and bring the paper back for a second weighing. This way your difference will be the mass of the chemical actually transferred, unaffected by any chemical that may have remained on the paper unnoticed. In this method you should use a hard, smooth paper—waxed paper is best—rather than coarse paper, such as paper towel, which is certain to trap powders and tiny crystals.

Laboratory balances are subject to corrosion. Both the balances and the balance area should be kept clean, and spilled chemicals should be cleaned up immediately.

Here are a few miscellaneous pointers on proper balance operation, given as a series of "do's and don'ts," with some items in both lists for emphasis:

DO
- Allow hot objects to cool to room temperature before weighing.
- Close the side doors or hood of a milligram balance while weighing.
- Record all digits allowed by the accuracy of the balance used, even if the last digit happens to be a zero on the right side of the decimal point.

DON'T
- Weigh objects that are warm or hot.
- Weigh objects that are wet (evaporation of water will change the mass).
- Weigh volatile liquids in uncovered vessels.
- Touch the object with your hand if you are using a milligram or analytical balance; your fingerprints have weight, too!
- Forget to check the zero on a milligram balance after weighing.
- Forget to record the mass to as many digits as the accuracy of the balance allows—and no more.

LABORATORY BURNERS

The function of a laboratory burner is to provide an adjustable mixture of natural gas and oxygen (from the air) that may be burned to produce the kind of flame required for a specific purpose. As a group the burners are called Bunsen burners, although most burners used today are improvements over the original Bunsen design. All have the same general features, and the Tirrill burner described in Figure LP–5 is representative of the group.

Gas enters the barrel of the burner from the center of the base, controlled by a valve in the base. Air enters through an opening at the bottom of the barrel where it screws onto the base. The amount of air admitted is governed by the position of the barrel. When the barrel is screwed down, the air opening is small. This limits the amount of air, and for a given amount of gas it gives a mixture that has a high gas-to-air ratio—a "rich" mixture. If the barrel is unscrewed to admit a large quantity of air, the mixture has a low gas-to-air ratio—a "lean" mixture. By adjusting the amount of gas at the bottom, you control the size of the flame; and by adjusting the amount of air with the barrel, you control the type of flame produced.

If you burn a mixture with very little air—a very high gas-to-air ratio, or a very rich mixture—the flame will be yellow and not very hot. The yellow color is from unburned carbon, which is deposited as soot on the bottom of any vessel that is heated with such a flame. Increasing the amount of air causes the flame to become less yellow and more blue, and finally all blue. As still more air is introduced, the blue flame separates into parts, a light-blue inner cone and a darker outer cone. The hottest part of the flame is just above the tip of the bright blue cone. If too much air is introduced, the entire flame will "rise" and burn noisily above the burner barrel.

Occasionally, a burner will "strike back" and burn the mixture inside the barrel where the two components first meet. You usually become aware of this condition by the noise produced in burning. If this happens, shut the burner off briefly, and then relight it. Be careful, however, because the barrel of a burner that is striking back becomes *very hot.*

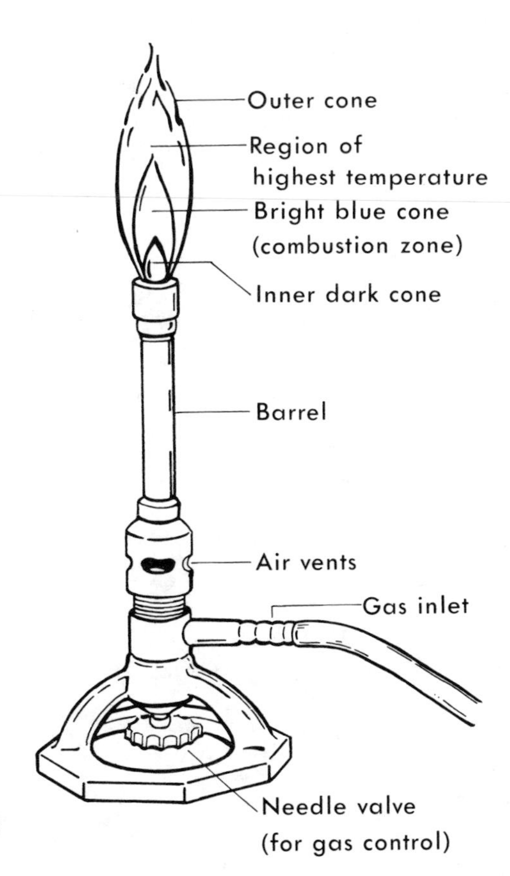

LP–5. Laboratory burner (Tirrill type) and flame.

If you are not familiar with laboratory burners, it is recommended that you light one and experiment with the various adjustments to see how they work. Don't be afraid of a burner. It is a simple device that cannot hurt you unless you put your hand in the flame or touch the barrel of a burner that has been striking back. It is also a rugged device that you will not damage without trying to.

The proper lighting procedure is to strike the match, open the gas valve at the laboratory desk completely, and then move the match flame to the burner just below the tip of the barrel, letting the top of the flame creep over the top of the barrel to light the gas. If your burner has no gas control valve in its base, you will have to control the gas at the desk; otherwise the desk valve is opened fully, and the gas flow is adjusted at the burner.

When you first use a burner to heat a cold object, start with a blue flame, but one that it is not very strong, or the object may crack. A blue flame that has no inner cone is ideal. After about a minute you can increase the amount of air in the flame to produce an inner cone and higher temperature. Crucibles and, with special precautions, test tubes may be heated directly in the flame. Crucibles are usually mounted on a clay triangle directly over the flame, and test tubes are held by hand in test-tube clamps. When you are heating a liquid in a beaker or flask, the vessel should be placed on a wire screen with an asbestos center, which is mounted on a ring stand or tripod. It is permissible to heat beakers and flasks made of Pyrex or other heat-resistant glass; but never, *under any circumstances,* heat calibrated volumetric glassware, such as graduated cylinders. There are two reasons for this "don't." First, the glass is not heat resistant and will probably crack. Second, heating would expand the glass and probably destroy the accuracy of the calibration.

PLACING RUBBER STOPPERS ON GLASS TUBING OR ROD

This book contains no experiment in which it is necessary to place a glass tube or thermometer through a rubber stopper to produce an airtight fit. We will therefore not describe the precautions that must be observed in this procedure, but rather describe the alternative procedure that may be used if the fit does not have to be airtight. A rubber stopper with one or more holes is cut from the side to one of the holes. The split stopper may then be held open, as shown in Figure LP–6, and the tubing or thermometer slipped into place. When the stopper is released it closes around the glass, holding it firmly in place.

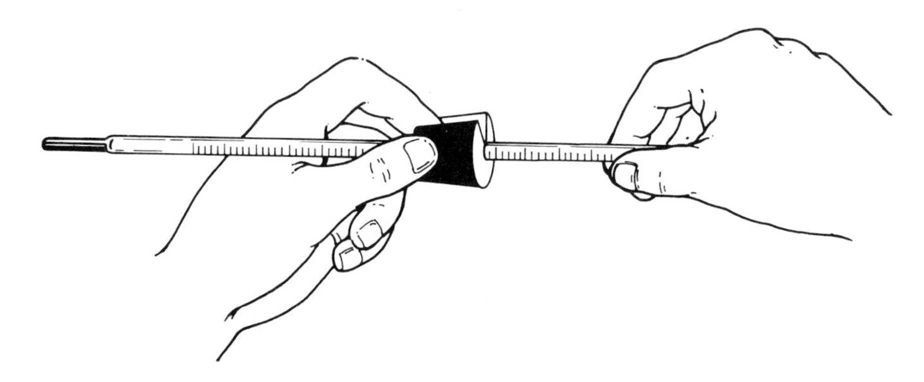

LP–6. The use of a split stopper.

Experiment 1Ⓜ

Properties and Changes of Matter

Performance Goals

1–1 Determine experimentally the solubility of a pure substance in a given liquid, or, in the case of two liquids, determine their miscibility.

1–2 Determine experimentally which of two immiscible liquids is more dense.

1–3 Determine whether or not a chemical reaction occurs when you combine two solutions, and state the evidence for your decision.

CHEMICAL OVERVIEW

All material things that compose our universe are referred to as **matter.** Matter is commonly defined as that which has mass and occupies space. In this experiment you will examine some of the characteristics of matter and be introduced to some of the language of science in which these characteristics are described.

A **pure substance** is a sample of matter that has identical properties throughout, and a definite, fixed composition. **Physical properties** are those characteristics of a substance that can be observed without changing the composition of the substance. Common physical properties are taste, color, odor, melting and boiling points, solubility, and density. **Chemical properties** describe the behavior of a substance when it changes its composition by reacting with other substances or decomposing into two or more other pure substances. The ability to burn and the ability to react with water are chemical properties.

Matter can undergo two types of changes, physical and chemical. **Physical changes** do not cause a change in composition, only in appearance. For example, when copper is melted, only a change of state occurs; no new substance is formed. In a **chemical change,** substances are converted into new products having properties and compositions that are entirely different from those of the starting materials. Wood, for example, undergoes a chemical change when it burns by reacting with oxygen in the air, forming carbon dioxide and water vapor as the new products.

When two liquids are mixed, the mixture may be completely uniform in appearance. In this case the liquids are said to be **miscible.** Some liquids are miscible in all proportions, while others have a limited range of miscibility. If the two liquids are not at all miscible, i.e., **immiscible,** two distinct layers will form when they are poured together. The liquid having the lower density will "float" on top of the other.

When a solid is added to and dissolves in a liquid, it is **soluble** in that liquid. The mixture formed is called a **solution.** A liquid solution is always clear; it may be colorless, or it may have a characteristic color. If the solid does not dissolve, it is said to be **insoluble.**

When two solutions are combined, there may be a chemical change, or **reaction,** in which new products form. If so, it will be evidenced by one of several visible changes. Among them are:

1. *Formation of a* **precipitate,** *or a solid product.* A precipitate is often very finely divided and distributed throughout the solution, giving a "cloudy" appearance. If allowed to stand, the precipitate will settle to the bottom of its container. The precipitate may be separated from the liquid by passing the mixture through a *filter* that collects the solid particles, but permits the solution to pass through.

2. *Formation of a gaseous product.* The gas produced bubbles out of the solution, a process called **effervescence.**
3. *Occurrence of a color change.* Usually a color change indicates the formation of a product with a color not originally present among the reactants. Sometimes the color will be the same as that of one of the reactants, but a darker or lighter shade.

In many cases no reaction occurs when two solutions are brought together.

SAFETY PRECAUTIONS AND DISPOSAL METHODS

Fumes from trichloroethane, xylene, and ammonia solutions are potentially harmful. Confine your use of these liquids to the fume hoods. Skin contact with these three liquids, or with hydrochloric acid, should be avoided. If it occurs, rinse the affected area thoroughly with water, and then wash with soap and water. Be sure to wear approved eye protection throughout the experiment.

Trichloroethane and xylene mixtures should be collected in stoppered bottles. **Do not pour them down the drain.** Solutions containing heavy metal precipitates should be collected in a separate container.

PROCEDURE

1. MIXING LIQUIDS

A) Place about 20 drops of trichloroethane into a small-size test tube. Note this approximate quantity, as you will have several occasions in the experiment to estimate this volume in a test tube. Add about 10 drops of water and gently shake the test tube, or mix the contents with a stirring rod. Are the two liquids miscible? Record your observation on the report sheet.

If the two liquids are *not* miscible, identify the liquid, trichloroethane or water, that is on top. You may determine which liquid is on top by the relative quantities placed into the test tube: you added two times as much trichloroethane as you did water. Record the name of the top liquid on your report sheet.

B) Discard the mixture from Part 1A and repeat the experiment, this time using about 20 drops of water first, followed by 10 drops of trichloroethane. If the liquids are not miscible, again record on the report sheet which liquid is on top.

C) Using a clean test tube, or the original one thoroughly rinsed with water, repeat the procedure with methanol (methyl alcohol) and water. It is not necessary this time or hereafter to reverse the order of adding liquids, as in Parts 1A and 1B. Again record your observations and conclusions.

D) Using a clean and thoroughly rinsed test tube, repeat the procedure again, this time with water and xylene. Record your observations and conclusions as before.

E) Using a clean *and dry* test tube (there must be no water present), perform the experiment once again, now using trichloroethane and xylene. Record your observations.

2. DISSOLVING A SOLID IN A LIQUID

In this part of the experiment and the next, you will be preparing solutions. The procedure is to take about 4 mm—just over 1/8 inch—of the solid on the tip of a spatula and place it into about 2 mL of deionized (or distilled) water in a test tube. Shake the test tube gently, or stir the contents with a clean, dry stirring rod. If none of the solid appears to dissolve, the substance is insoluble. If *any* of it dissolves, but a small amount does not, add more water to get all of the solid into the solution.

A) Place a small quantity of barium chloride, $BaCl_2$, in water as described above. Does the solid dissolve in the water? Record your observations and save the mixture for further use.

B) Repeat the above procedure with sodium sulfate, Na_2SO_4. Record your observations and save the solution.

C) Combine the contents of the test tubes from Steps 2A and 2B. Record your observations. Set the test tube aside for 5 to 10 minutes and examine it again. Record what you see.

D) Repeat the procedure in Step 2A, this time using barium sulfate, $BaSO_4$. Record your observations.

3. MIXING SOLUTIONS

A) Prepare and mix solutions of iron(III) chloride, $FeCl_3$, and potassium thiocyanate, KSCN, as in Steps 2A–2C. Has a chemical reaction taken place? Whether your answer is yes or no, record the evidence on which it is based.

B) In like manner, prepare and mix solutions of sodium chloride, NaCl, and ammonium nitrate, NH_4NO_3. Record your observations and identify evidence that a reaction has or has not occurred.

C) Prepare a solution of sodium carbonate, Na_2CO_3 (or potassium carbonate, K_2CO_3). Using an eye dropper, add 2 or 3 drops of hydrochloric acid, HCl, watching carefully for any evidence of a reaction. Then discharge the remaining contents of the eye dropper into the solution, and again watch for a reaction. Record your observations.

D) Repeat the procedure of Step 3A, using calcium chloride, $CaCl_2$ and sodium carbonate, Na_2CO_3. Record your observations.

E) Prepare a solution of copper(II) sulfate, $CuSO_4$, and add concentrated ammonia solution, $NH_3(aq)$, to it a drop at a time. (Ammonia solutions are often labeled NH_4OH.)

Experiment 1
Work Page

NAME ....

DATE .... SECTION ....

PART 1— MIXING LIQUIDS

Step	*Mixture*	*Miscible or Immiscible*	*Liquid on Top*	*More Dense Liquid*
1A	Trichloroethane (20 drops) Water (10 drops)			
1B	Water (20 drops) Trichloroethane (10 drops)			
1C	Water Methanol			
1D	Water Xylene			
1E	Trichloroethane Xylene			

QUESTIONS

1) Is it possible from Parts 1A–1D to determine which of the liquids, xylene or trichloroethane, is more dense? If so, identify the liquid with the greater density and explain how you reached your conclusion; if not, explain why.

More dense liquid: ____

Explanation:

2) Is it possible from Part 1E alone to determine which of the liquids, xylene or trichloroethane, is more dense? If so, identify the liquid with the greater density and explain how you reached your conclusion; if not, explain why.

More dense liquid: ____

Explanation:

PART 2 — DISSOLVING A SOLID IN A LIQUID

2A) Barium chloride: Soluble (____________) or insoluble (____________)?

2B) Sodium sulfate: Soluble (____________) or insoluble (____________)?

2C) Mixture of contents of test tubes from 2A and 2B:

Immediate appearance:

Appearance 5–10 minutes later:

2D) Barium sulfate: Soluble (____________) or insoluble (____________)?

3) Based on your observations from Steps 2A, 2B, and 2D, suggest an explanation for your observation in Step 2C.

PART 3 — MIXING SOLUTIONS

Step	*Solutions Combined*	*Reaction: Yes or No*	*Evidence*
3A	Iron(III) chloride Potassium thiocyanate		
3B	Sodium chloride Ammonium nitrate		
3C	Sodium carbonate Hydrochloric acid		
3D	Calcium chloride Sodium carbonate		
3E	Copper(II) sulfate Ammonia solution		

Experiment 1

Report Sheet

NAME ....

DATE .... SECTION ....

PART 1 — MIXING LIQUIDS

Step	*Mixture*	*Miscible or Immiscible*	*Liquid on Top*	*More Dense Liquid*
1A	Trichloroethane (20 drops) Water (10 drops)			
1B	Water (20 drops) Trichloroethane (10 drops)			
1C	Water Methanol			
1D	Water Xylene			
1E	Trichloroethane Xylene			

QUESTIONS

1) Is it possible from Parts 1A–1D to determine which of the liquids, xylene or trichloroethane, is more dense? If so, identify the liquid with the greater density and explain how you reached your conclusion; if not, explain why.

   More dense liquid: ____________________

   Explanation:

2) Is it possible from Part 1E alone to determine which of the liquids, xylene or trichloroethane, is more dense? If so, identify the liquid with the greater density and explain how you reached your conclusion; if not, explain why.

   More dense liquid: ____________________

   Explanation:

PART 2 — DISSOLVING A SOLID IN A LIQUID

2A) Barium chloride: Soluble (____________) or insoluble (____________)?

2B) Sodium sulfate: Soluble (____________) or insoluble (____________)?

2C) Mixture of contents of test tubes from 2A and 2B:

Immediate appearance:

Appearance 5–10 minutes later:

2D) Barium sulfate: Soluble (____________) or insoluble (____________)?

3) Based on your observations from Steps 2A, 2B, and 2D, suggest an explanation for your observation in Step 2C.

PART 3 — MIXING SOLUTIONS

Step	*Solutions Combined*	*Reaction: Yes or No*	*Evidence*
3A	Iron(III) chloride Potassium thiocyanate		
3B	Sodium chloride Ammonium nitrate		
3C	Sodium carbonate Hydrochloric acid		
3D	Calcium chloride Sodium carbonate		
3E	Copper(II) sulfate Ammonia solution		

Experiment 1
Advance Study Assignment

NAME ....

DATE .... SECTION ....

1) Distinguish between physical and chemical properties. Give an example of each.

2) Classify each of the following as a physical or chemical change:

a) Iron rusting ______________ c) Burning paper _________________

b) Boiling water _____________ d) Acquiring a suntan _____________

3) Identify three forms of evidence that a chemical reaction has occurred:

a)

b)

c)

Experiment 2

Measured Quantities and Derived Quantities: Significant Figures

Performance Goals

2–1 Using devices available in the laboratory, make length, mass, and volume measurements of given samples of matter.

2–2 Add, subtract, multiply, and divide laboratory measurements and express the result in the proper number of significant figures.

CHEMICAL OVERVIEW

Significant Figures. A chemist makes many measurements in the laboratory, using tools that are sometimes very simple and sometimes very complicated, depending on the "exactness" desired. No measurement is really *exact.* Every measurement is limited by some **uncertainty,** a plus-or-minus range that must be attached to the measured quantity if you are to be sure that the true value is stated by the measurement. This uncertainty arises from the measurement process or from the measuring instrument. By using an instrument capable of finer measurements we can reduce the uncertainty; but we can never eliminate it entirely. We will use **significant figures** to express uncertainty in a measured quantity, as well as any *derived quantity* that may result from calculations based on measured quantities.

Your textbook probably discusses significant figures in greater detail than we can here. You should read that material as part of your preparation for this experiment. The main ideas behind significant figures as they will be used in this laboratory manual are summarized as follows:

1. The concept of significant figures applies to *measured quantities* because of the uncertainty associated with every measurement. It does *not* apply to exact numbers, such as counting numbers or numbers that are exact by definition.
2. The number of significant figures in a measurement is the number of figures that are known accurately plus one that is doubtful. The doubtful digit in a number properly expressed in significant figures is the *last digit shown.*
3. Counting significant figures always begins with the first nonzero digit and ends with the last digit shown — the doubtful digit. Notice that there is *no reference to the decimal point* in this statement. The location of the decimal point is determined by the units in which a measurement is expressed. It has nothing to do with the measurement process, and therefore nothing to do with the number of significant figures in a quantity.
4. Tail-end zeros to the right of the decimal point are used to indicate that they are significant, in accordance with Number 2 above. Accordingly, a tail-end zero *must* be used to the right of the decimal if it is the doubtful digit.
5. Tail-end zeros to the left of the decimal in large numbers are necessary to show the magnitude of the number, but they may or may not be significant. Use exponential notation so any doubtful digit zero appears to the right of the decimal as the last digit shown, in accordance with Numbers 2 and 4 above.

6. Rounding off: If the first digit to be dropped is less than 5, leave the preceding digit unchanged; if the first digit to be dropped is 5 or more, increase the preceding digit by 1. (If your instructor prefers a different rule for rounding off, use it. Only the doubtful digit will be affected.)
7. Addition–subtraction rule: Round off the answer to the first column that has a doubtful digit.
8. Multiplication–division rule: Round off the answer to the *same number* of significant figures as the *smallest* number of significant figures in any factor.

Mass Measurement. The main laboratory objective of this experiment is to give you practice in the proper use of balances. You will be given a set of objects to weigh on the balances you will use throughout the semester. The masses of two of these objects have been determined in advance, and are listed on a separate sheet or posted in the laboratory. By comparing your mass measurements with those on the list you may assure yourself that you are using the balance correctly — assuming, of course, that your measurement is the same as the mass on the list. The masses listed are subject to the same laboratory uncertainty as your measurements, so you probably will not duplicate those values exactly. If your measurement on a centigram balance is within 0.04 gram of the listed mass, or within 0.004 gram on a milligram balance, you may regard your values as acceptable. If you do not reach this agreement, and cannot determine a cause for the discrepancy, ask an assistant or the instructor for help.

In finding the mass of a chemical, you **never** place the chemical directly on the pan of a balance. Instead you follow a procedure known as **weighing by difference.** Because this method will be referred to in many experiments in this book, it is described in the Laboratory Procedures section (see page 8) rather than here. Please read that material now as part of this overview.

Length and Area Measurements. At one point in this experiment you are asked to find the areas of two geometric figures, a rectangle and a triangle. You will use a typical ruler to find the required length measurements. The area of a rectangle is the product of the length l multiplied by the width w:

$$A = l \times w \tag{2.1}$$

The area of a triangle is one-half the product of the length of any side l and the height h drawn to that side:

$$A = \tfrac{1}{2}(l \times h) \tag{2.2}$$

The height of a triangle to any side is the perpendicular distance between that side and the opposite vertex. In the illustration, h is the height of the triangle drawn to side AB:

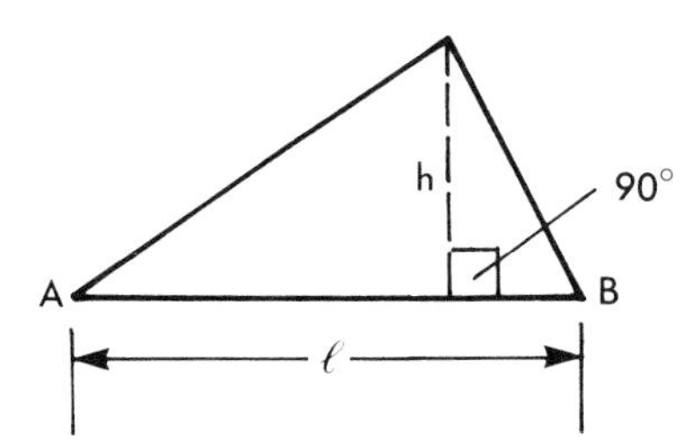

PROCEDURE

1. MEASUREMENT OF MASS

Obtain a set of three solid objects, two of known mass and one of unknown mass. Record the set number on your data sheet — *now.* Measure and record the mass of each object on each of the balances assigned to you. From the list provided, also write the predetermined masses of objects A and B. Compare these values with those you obtained. If your measurements are not within acceptable range (0.04 g on a centigram balance and 0.004 g on a milligram balance) of the listed mass, recheck your weighings. Ask for help if needed — this is the time to learn how to use the balances properly.

Now, using any one balance you choose, determine the masses of your three objects in all possible combinations of two — (A + B), (B + C), and (A + C) — and the total mass of the three pieces weighed together (A + B + C). After these values are recorded, complete the addition exercises that follow on the report sheet, and answer the questions. For individual masses of A, B, and C, use the values obtained on the same balance used for weighing your combinations. These exercises should prove the well-known mathematical and physical fact that the whole is equal to the sum of its parts — or is it?

2. DETERMINATION OF MASS AND VOLUME OF A SAMPLE OF WATER

Measure the mass of a clean, dry, 25- or 50-mL graduated cylinder on a centigram balance. Remove the cylinder from the balance and pour into it 15 to 20 mL of water. Do not attempt to make the volume come out to some whole number of milliliters; this would defeat the purpose of the exercise. Estimate the volume of the water to the nearest 0.1 mL, being careful to read the *bottom* of the meniscus at eye level. (If you are not familiar with reading volumetric glassware, see the discussion of this topic on page 7 in the Laboratory Procedures section.) Be sure the outside of the cylinder is dry, and then measure the mass of the cylinder and water. Record all measurements on your data sheet.

3. DETERMINATION OF AREA — A DERIVED QUANTITY

A) With one exception — this one — all measurements called for in this laboratory manual will be made in metric units. In this example you will make identical measurements in both the English and metric systems in order that you may compare them for convenience in use. You are to find the area of rectangle ABCD on page 27. Measure and record to the nearest sixteenth of an inch both the length and width of the rectangle. Now, using those measurements just as you have recorded them — and that includes the fractions, *not their decimal equivalents* — calculate the area of the rectangle in square inches. Note that, in using the fractions of English measurements, you must calculate the area by longhand arithmetic; calculators cannot be used with ordinary fractions. Finally, convert your square-inch area to the equivalent number of square feet. Show all calculations in the space provided.

B) Repeat the above measurements, this time recording the length and width of the rectangle in centimeters, measured to the nearest 0.1 cm. Calculate the area in square centimeters, and then convert that area to square meters. While a calculator could be used with metric measurements, you should perform this calculation by longhand arithmetic too, so your comparison of English and metric calculations is not biased by an electronic tool. Record the results for both areas, and then show the areas rounded off to the proper number of significant figures.

C) Measure the lengths of the three sides of triangle XYZ on page 28 and record the measurements to the nearest 0.1 cm. Then draw, measure, and record the three heights of the triangle, one to each of the three sides. Finally, for each length and height measurement, calculate the area of the triangle by Equation 2.2. Record your results, first with the full calculator readout or longhand arithmetic product, and then rounded off to the proper number of significant figures.

Experiment 2
Work Page

NAME ....

DATE .... SECTION ....

PART 1 — DATA

Balance					
Object A					
Object B					
Object C					
Objects (A + B)					
Objects (B + C)					
Objects (A + C)					
Objects (A + B + C)					

Mass set number ______________ Listed mass of A ______________g; of B ______________g

PART 1 — RESULTS

A (from table above): ______________g

B (from table above): ______________g

Sum: A + B: ______________g

Observed (A + B): ______________g

Difference: ______________g

C (from table above): ______________g

A (from table above): ______________g

Sum: A + C: ______________g

Observed (A + C): ______________g

Difference: ______________g

B (from table above): ______________g

C (from table above): ______________g

Sum: B + C: ______________g

Observed (B + C): ______________g

Difference: ______________g

A (from table above): ______________g

B (from table above): ______________g

C (from table above): ______________g

Sum: A + B + C: ______________g

Observed (A + B + C): ______________g

Difference: ______________g

PART 1 — QUESTION

Probably there are one or more examples among your measurements in which the observed and calculated values for the same mass do not match. Account for this fact, or for this expectation in the event you had zero difference for all four examples.

PART 2 — DATA, RESULTS, AND QUESTIONS

Mass of cylinder: ____________________ g

Volume of water: ____________________ mL

Mass of cylinder + water: ____________________ g

Mass of water (show calculation): ____ g

1) The mass of 1 mL of water is presumably 1 g. Account for the fact that the milliliters of water and grams of water are not identical numbers.

2) Which instrument do you think gives a more accurate estimate of the quantity of water in a sample, a centigram balance or the graduated cylinder you used? Explain.

Experiment 2
Work Page

NAME ....

DATE .... SECTION ....

PART 3 — RECTANGLE: FIGURE, DATA, AND RESULTS

Dimension	*English Units*	*Metric Units*
Length	in	cm
Width	in	cm
Area	in^2	Full readout: cm^2
		Rounded off: cm^2
	ft^2	Full readout: m^2
		Rounded off: m^2

CALCULATIONS

PART 3 — TRIANGLE: FIGURE, DATA, AND RESULTS

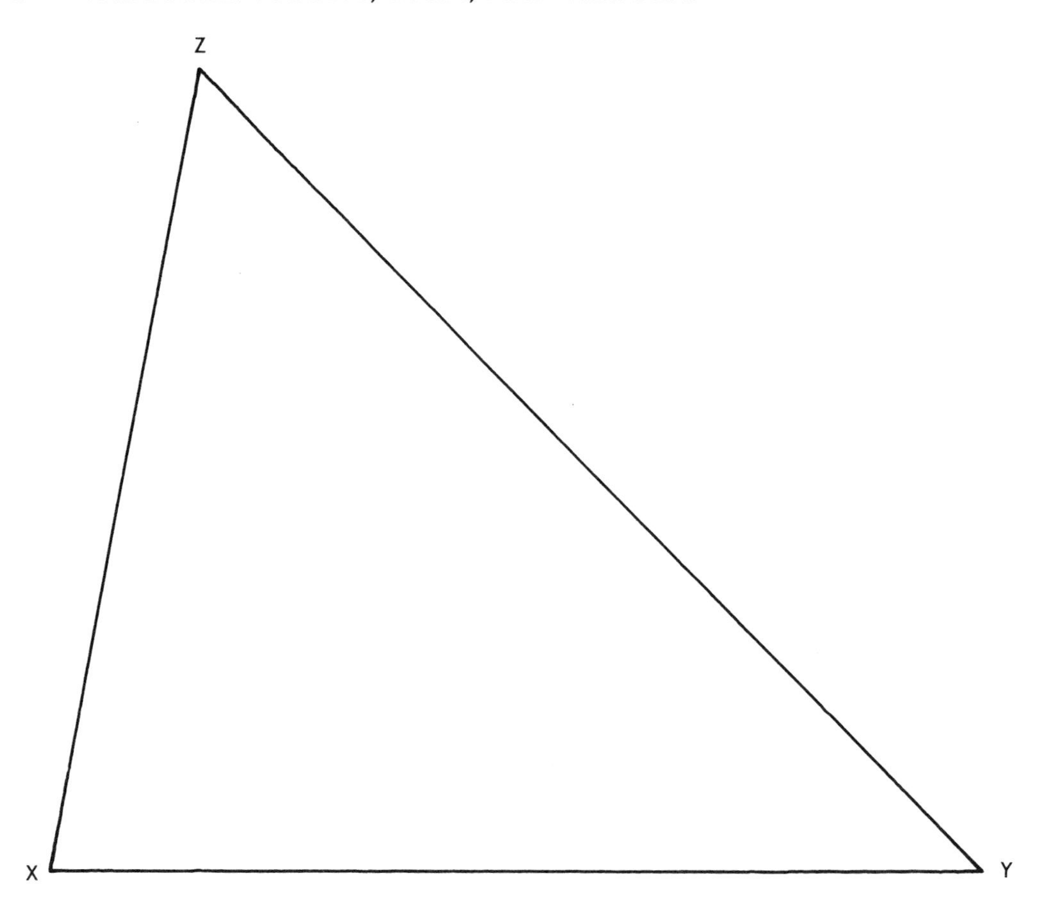

Side		$\overline{XY}$	$\overline{YZ}$	$\overline{ZX}$
Length, cm				
Height, cm				
Area, cm^2	Full calculator readout or longhand calculation			
	Corrected to proper number of significant figures			

CALCULATIONS

Show at least one area calculation completely.

Experiment 2

Work Page

NAME ....

DATE .... SECTION ....

PART 3 — QUESTIONS

1) In which units — inches or centimeters — are the dimensions of Figure ABCD more easily measured and the area calculated? Why?

2) Which value for a derived quantity do you think is more reliable, a full calculator readout, or a value corrected to the proper number of significant figures? Explain in relation to the three areas you calculated for the same triangle above.

Experiment 2
Report Sheet

NAME ....

DATE .... SECTION ....

PART 1 — DATA

Balance					
Object A					
Object B					
Object C					
Objects (A + B)					
Objects (B + C)					
Objects (A + C)					
Objects (A + B + C)					

Mass set number ______________ Listed mass of A ______________g; of B ______________g

PART 1 — RESULTS

A (from table above): ______________g

B (from table above): ______________g

Sum: A + B: ______________g

Observed (A + B): ______________g

Difference: ______________g

C (from table above): ______________g

A (from table above): ______________g

Sum: A + C: ______________g

Observed (A + C): ______________g

Difference: ______________g

B (from table above): ______________g

C (from table above): ______________g

Sum: B + C: ______________g

Observed (B + C): ______________g

Difference: ______________g

A (from table above): ______________g

B (from table above): ______________g

C (from table above): ______________g

Sum: A + B + C: ______________g

Observed (A + B + C): ______________g

Difference: ______________g

PART 1 — QUESTION

Probably there are one or more examples among your measurements in which the observed and calculated values for the same mass do not match. Account for this fact, or for this expectation in the event you had zero difference for all four examples.

PART 2 — DATA, RESULTS, AND QUESTIONS

Mass of cylinder: ____________________ g

Volume of water: ____________________ mL

Mass of cylinder + water: ____________________ g

Mass of water (show calculation): ____ g

1) The mass of 1 mL of water is presumably 1 g. Account for the fact that the milliliters of water and grams of water are not identical numbers.

2) Which instrument do you think gives a more accurate estimate of the quantity of water in a sample, a centigram balance or the graduated cylinder you used? Explain.

Experiment 2

Report Sheet

NAME ....

DATE .... SECTION ....

PART 3 — RECTANGLE: FIGURE, DATA, AND RESULTS

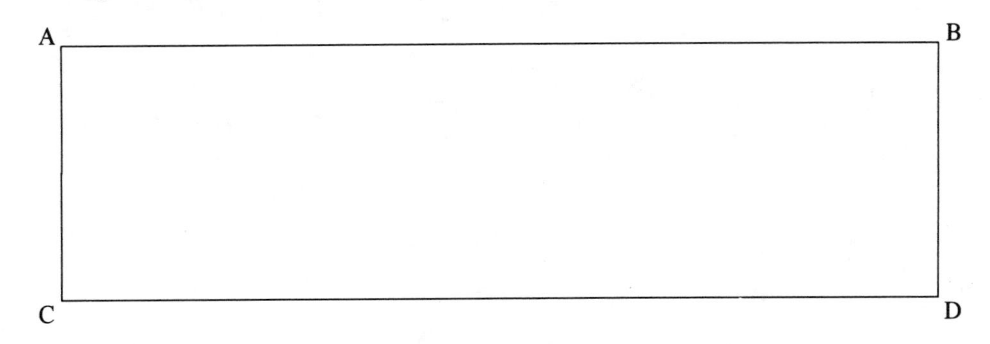

Dimension	*English Units*	*Metric Units*
Length	in	cm
Width	in	cm
Area	in^2	Full readout: cm^2
		Rounded off: cm^2
	ft^2	Full readout: m^2
		Rounded off: m^2

CALCULATIONS

PART 3 — TRIANGLE: FIGURE, DATA, AND RESULTS

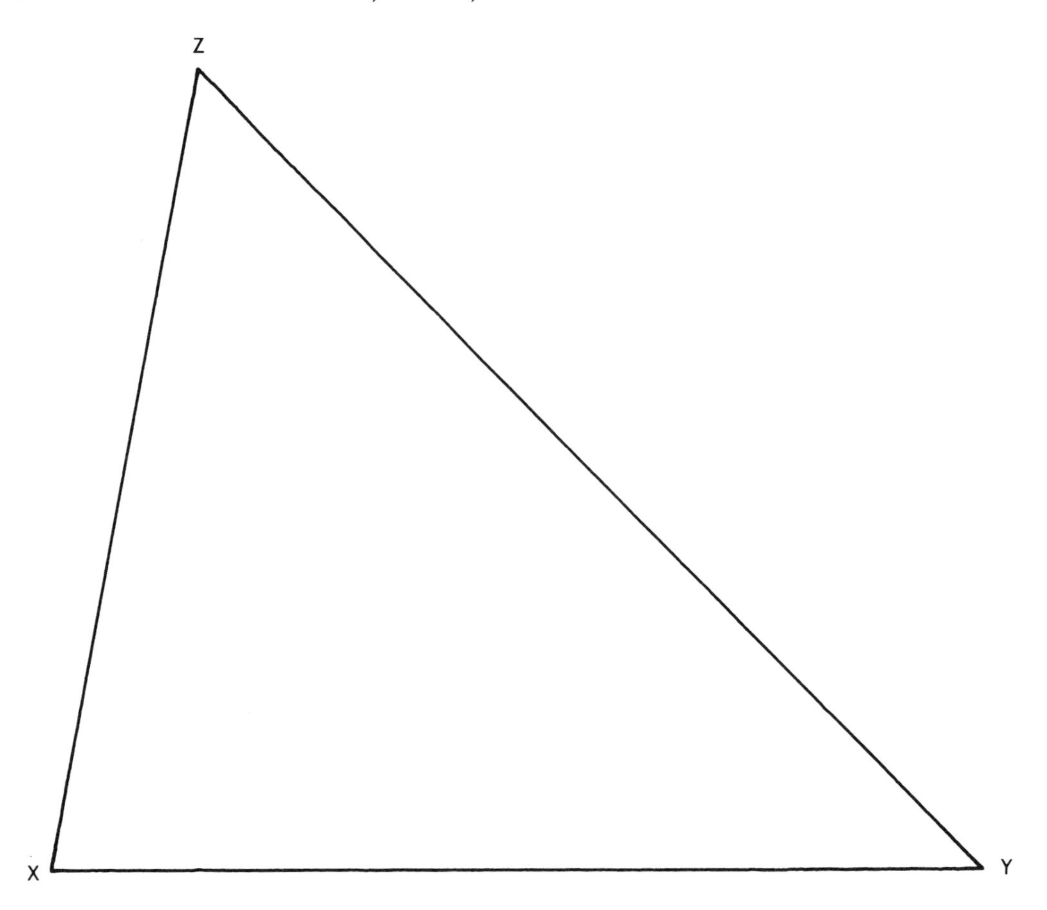

Side		$\overline{XY}$	$\overline{YZ}$	$\overline{ZX}$
Length, cm				
Height, cm				
Area, cm^2	Full calculator readout or longhand calculation			
	Corrected to proper number of significant figures			

CALCULATIONS

Show at least one area calculation completely.

Experiment 2

Report Sheet

NAME ....

DATE .... SECTION ....

PART 3 — QUESTIONS

1) In which units — inches or centimeters — are the dimensions of Figure ABCD more easily measured and the area calculated? Why?

2) Which value for a derived quantity do you think is more reliable, a full calculator readout, or a value corrected to the proper number of significant figures? Explain in relation to the three areas you calculated for the same triangle above.

Experiment 2

NAME ....

Advance Study Assignment

DATE .... SECTION ....

1) Explain briefly what is meant by "a centigram balance."

2) Why do we work with significant figures — or, in advanced courses, even more sophisticated procedures — when calculating with or analyzing laboratory measurements?

3) How many significant figures are in the measurement 0.0350 meters? ________________________

4) Convert the measurement 62.05 meters to millimeters, and express the result in the proper number of significant figures.

Experiment 3

Calibration of a Thermometer

Performance Goals

3–1 Calibrate a thermometer by measuring melting or freezing points of pure substances.
3–2 Construct a calibration curve for a thermometer.

CHEMICAL OVERVIEW

More than most laboratory devices, thermometers need to be calibrated if the measured temperatures are to be considered reliable. The inaccuracy of a thermometer may be due to several causes, including irregularity in the capillary tube and thermal expansion and contraction during temperature cycling. Thermometers for precision laboratory work may be calibrated against a very highly accurate thermometer available from the National Bureau of Standards. Alternately, a thermometer can be calibrated experimentally by matching its readings with the known melting or freezing points of pure substances. The second method will be used in this experiment.

The melting point (or freezing point), a physical property of pure substances, is the temperature at which a substance changes between the solid and the liquid states. The melting behavior of a compound is often a simple test of its purity. An impure compound melts over a range of temperatures that are lower than the melting point of the pure compound.

In this experiment, calibration will be carried out by checking the accuracy of your thermometer at the following temperatures:

1. Melting point of ice, 0.0°C.
2. Freezing point of glacial acetic acid, 16.7°C.
3. Melting point of dodecanal (lauraldehyde), 44.5°C.
4. Melting point of naphthalene, 80.2°C.
5. Melting point of α-naphthol, 93.4°C.

Some thermometers are designed for total immersion in the temperature zone being measured; others have an immersion line etched on them. In both cases, if you use the instrument as intended, you will probably minimize temperature errors. For best results, the calibration of either type of thermometer will be more accurate if it is carried out under the conditions of its intended use. Since most work with the thermometer in the general chemistry laboratory is performed when only the lower portion of the thermometer is immersed in the test zone, it is advisable to calibrate your thermometer for partial immersion. If the thermometer has no immersion line etched on it, choose a fixed position on the thermometer about 6 to 7 cm above the bulb and always immerse the thermometer to that depth.

SAFETY PRECAUTIONS AND DISPOSAL METHODS

Do not force a thermometer into a bored rubber stopper. Instead, use a split rubber stopper as described on page 10. When handling organic chemicals, use a spatula. Avoid skin contact. Glacial acetic acid vapors are very potent, and should not be breathed.

Dispose of glacial acetic acid in a waste container. Excess organic solids should be collected in a container.

PROCEDURE

1. MELTING POINT OF ICE

A) In a 250-mL beaker place more than enough crushed ice to cover the immersion line of the thermometer and add just enough deionized water to cover the ice.

B) Insert a thermometer into a split stopper (see page 10), and then close a clamp around the stopper to hold it, as shown in Figure 3–1. Place the thermometer in the slush to the immersion mark, taking care that the thermometer does not touch the bottom or the walls of the beaker. Stir the mixture with a stirring rod.

C) When the level of the mercury column remains constant for at least 1 minute, record the thermometer reading to the nearest 0.1°C. Enter your reading on the report sheet.

2. FREEZING POINT OF GLACIAL ACETIC ACID

A) Pour 15 mL of glacial acetic acid into a large, clean, dry test tube. Using a ring stand and a clamp, suspend the test tube in the beaker of ice used in Step 1 (see Figure 3–1). Be sure the level of the acetic acid in the test tube is below the level of the slush in the beaker. Replenish the ice if most of it has melted.

B) Lower the thermometer into the acetic acid until the immersion line is at the liquid surface.

C) Loop a circular wire stirrer around the thermometer in the test tube, and stir the acetic acid by raising and lowering the wire. Observe and record the temperature to the nearest 0.1°C when the first crystals appear and the temperature remains constant.

If the solution freezes very rapidly, you may not observe a constant temperature. If this happens, allow the acetic acid to thaw, and repeat the freezing step. Do not allow the whole mass of acetic acid to become solid; measure the temperature when both crystals and liquid exist together.

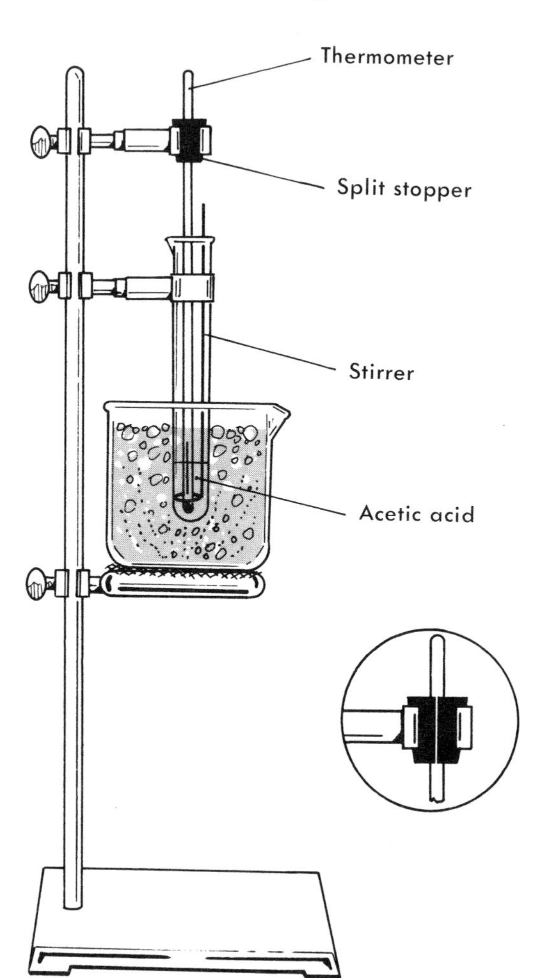

Figure 3–1. Apparatus for freezing point determination.

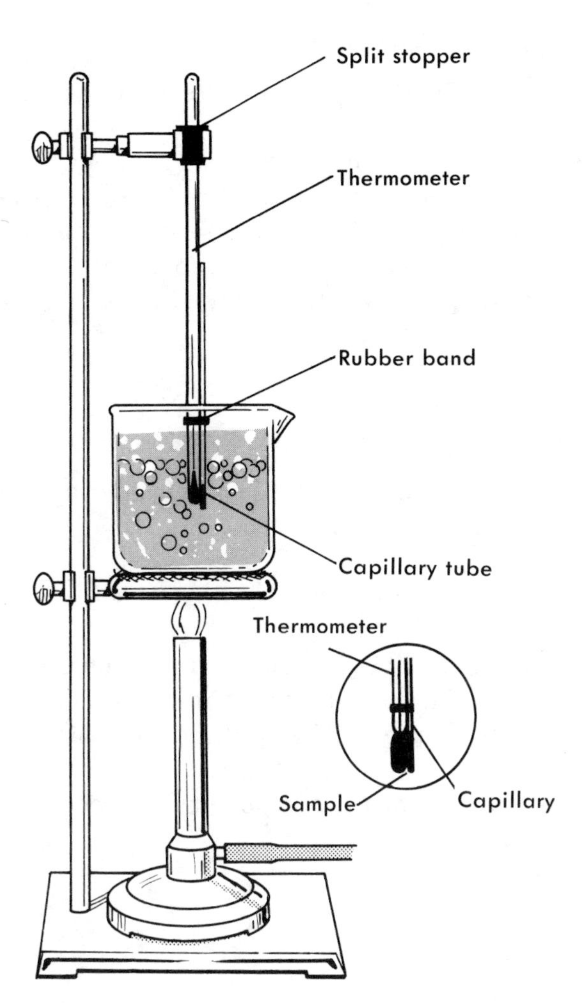

Figure 3–2. Apparatus for determination of melting point.

3. MELTING POINT OF LAURALDEHYDE

A) Crush a small amount of lauraldehyde on a watch glass with a spatula, and scrape the powder into a mound. Put the powder into a melting-point capillary by pushing its open end into the mound and then compacting the powder by allowing the vertically held capillary to drop a short distance onto a hard surface (such as a table top). Do not press too much powder into the capillary tube during each step, because the powder will pack near the opening and will not move to the bottom of the tube. Add material until you have a tightly packed column about three-fourths of the length of the thermometer bulb.

B) Fasten the capillary to the thermometer by means of a rubber band or a thin slice of rubber tubing, making sure the packed part of the capillary is level with the thermometer bulb.

C) Fill a 250-mL beaker with water and assemble the apparatus shown in Figure 3–2. The water level should be below the open end of the capillary tube and at the immersion level of the thermometer.

D) Heat the water slowly with constant stirring and carefully watch the capillary tube. Read the thermometer at the first appearance of a liquid and again when the last of the solid disappears. Record these temperatures. For the purpose of calculating the temperature correction, take the midpoint of the temperature range and compare it to the known melting point. For example, if the first liquid appears at 44.2°C, and the sample is fully melted at 44.6°C, then use 44.4°C as the temperature reading for the observed melting point.

4. MELTING POINT OF NAPHTHALENE

Using a fresh capillary, determine and record the melting point of naphthalene by the same method described in Step 3.

5. MELTING POINT OF α-NAPHTHOL

Repeat the procedure for Step 3 using α-naphthol in a clean capillary tube.

> **Note: It is possible to combine Steps 3 through 5 by measuring the melting points of the substances in a continuous fashion. To do so, prepare three capillary tubes, one for each compound listed in Steps 3, 4, and 5. Fasten all three capillaries to the thermometer at the same time. Proceed as described in Step 3, heating the water slowly. Observe the melting ranges of the compounds in the order of increasing temperatures.**

CALCULATIONS

In Steps 3 to 5, record the midpoints of the temperature ranges as the melting points. For all steps, calculate the temperature correction by the equation

$$\text{Correction} = \text{known value} - \text{observed value} \tag{3.1}$$

List the correction as positive if it must be added to the observed reading to get the known, but negative if the correction must be subtracted. (Example: Your thermometer is used to measure the freezing point of benzene, and the range is from 5.3 to 5.5°C, giving a midpoint of 5.4°C. The known freezing point is 5.5°C. Correction: 5.5°C − 5.4°C = +0.1°C.) Prepare a correction curve on the paper provided by plotting the corrections against the temperature reading. Draw a smooth curve through the points. Submit this curve with your report sheet.

Experiment 3

Work Page

NAME ....

DATE .... SECTION ....

Measurement	*Observed Melting Range*		*Observed Melting (Freezing) Point*	*Known Melting (Freezing) Point*	*Correction (Include sign)*
	Melting Begins	*Melting Complete*			
Melting of ice					
Freezing of acetic acid					
Melting of lauraldehyde					
Melting of naphthalene					
Melting of α-naphthol					

Immersion line: ________________________________ (cm from bottom of bulb).

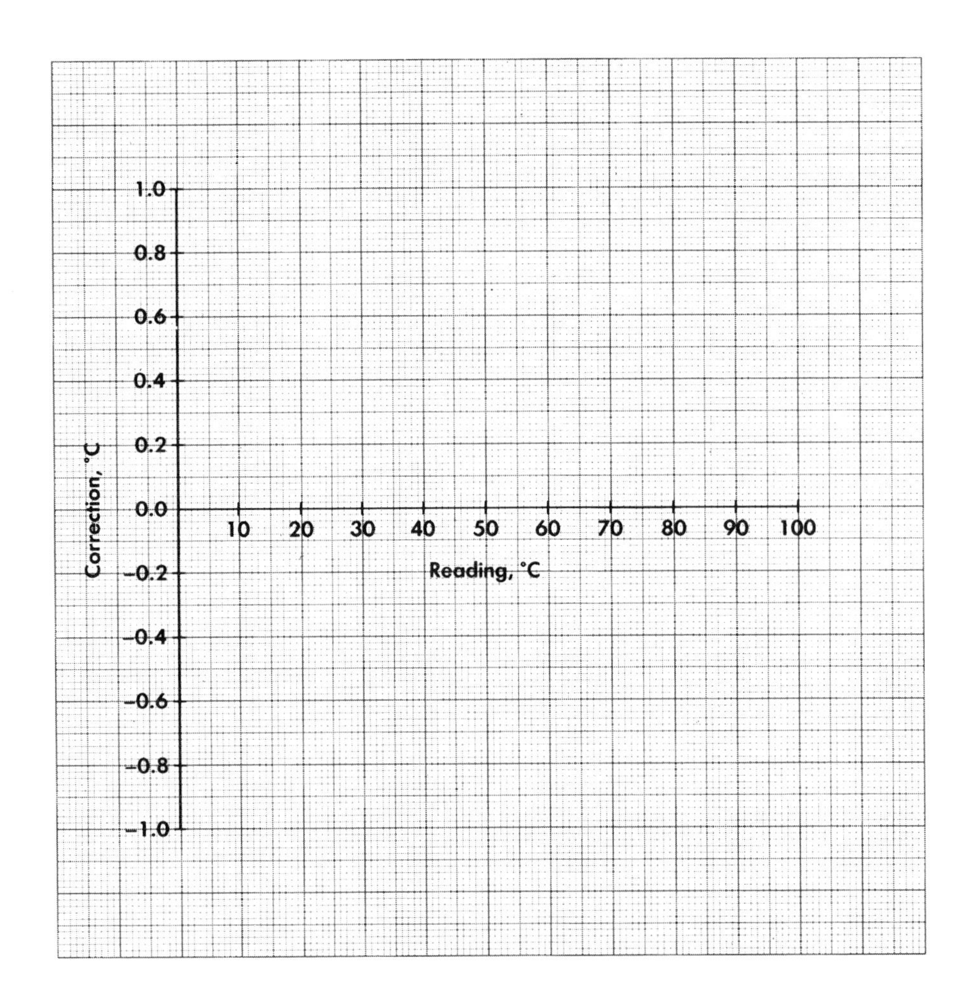

Calibration Curve of a Thermometer

Experiment 3

Report Sheet

NAME ....

DATE .... SECTION ....

Measurement	*Observed Melting Range*		*Observed Melting (Freezing) Point*	*Known Melting (Freezing) Point*	*Correction (Include sign)*
	Melting Begins	*Melting Complete*			
Melting of ice					
Freezing of acetic acid					
Melting of lauraldehyde					
Melting of naphthalene					
Melting of α-naphthol					

Immersion line: ________________________________ (cm from bottom of bulb).

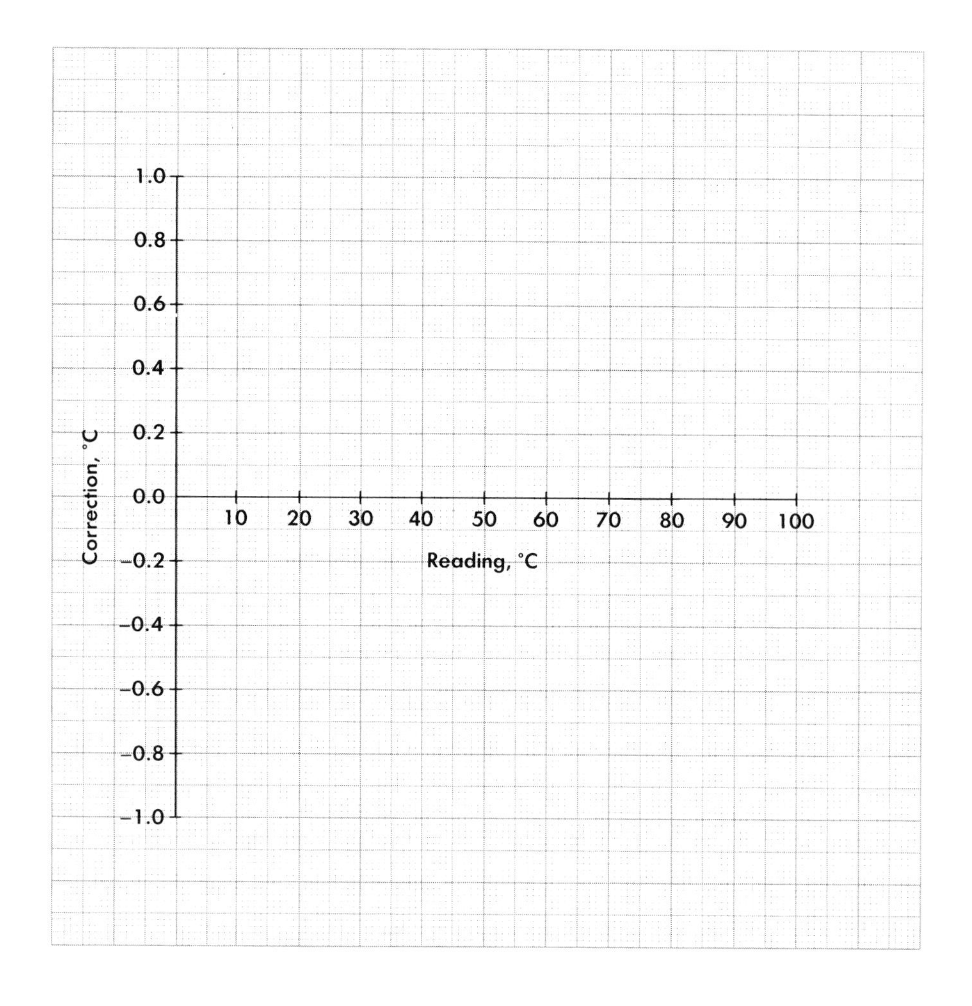

Calibration Curve of a Thermometer

Experiment 3

Advance Study Assignment

NAME ....

DATE .... SECTION ....

1) What error will occur in the calibration of the thermometer if a substance is contaminated?

2) Calculate the correction for a thermometer that measures the melting range of a compound as 89.6 to 89.8°C; the theoretical melting point is known to be 89.6°C.

3) What is the difference between the melting point of ice and the freezing point of water?

Experiment 4(M)

Separation of Cations by Paper Chromatography

Performance Goals

4–1 Separate a mixture of cations by paper chromatography and calculate their R_F values.
4–2 Analyze an unknown mixture of cations by paper chromatography.

CHEMICAL OVERVIEW

Chromatography, which means "the graphing of colors," obtains its name from the early experiments of Tswett, who, in 1906, succeeded in separating a mixture of colored pigments obtained from leaves. A solvent mixture, carrying the pigments, was allowed to pass through a glass column packed with chalk. At the end of the experiment, the pigments were separated in colored bands at various distances from the starting level. This method is now known as column chromatography.

Chromatography may now be applied to colorless compounds and to ions. Paper chromatography is a more recent and much faster separation technique than column chromatography. It may be used for the separation of substances by a solvent moving on sheets or strips of filter paper. The filter paper is referred to as the **stationary phase,** or **adsorbent.** The mixture of solvents used to carry the substances along the paper is called the **mobile phase,** or **solvent system.**

In practice, a sample of the solution containing the substances to be separated is dried on the paper. The end of the paper is dipped into the solvent system so that the separation sample is slightly above the liquid surface. As the solvent begins to soak the paper, rising by capillary action, it transports the sample mixture upward. Each component of the mixture being separated is held back by the stationary phase to a different extent. Also, each component has a different solubility in the mobile phase and therefore moves forward at a different speed. A combination of these effects causes each component of the mixture to progress at a different rate, resulting in separation.

In a given solvent system, using the same adsorbent at a fixed temperature, each substance can be characterized by a constant **retention factor,** R_F. By definition,

$$R_F = \frac{\text{Distance from origin to center of spot}}{\text{Distance from origin to solvent front}} \tag{4.1}$$

where the **origin** is the point at which the sample was originally placed on the paper and the **solvent front** is the line representing the most advanced penetration of the paper by the solvent system. The R_F value is a characteristic property of a species, just as the melting point is a characteristic property of a compound.

In this experiment you will separate a mixture of iron(III), copper(II), and cobalt(II) ions, Fe^{3+}, Cu^{2+}, and Co^{2+}, respectively. Each ion forms a different colored complex when sprayed with a solution containing potassium hexacyano ferrate(II), $K_4[Fe(CN)_6]$.

SAFETY PRECAUTIONS AND DISPOSAL METHODS

Acetone is *extremely flammable.* Its vapors can ignite even when the liquid is a considerable distance from an open flame, so be sure no such flame is operating in the vicinity of your work area. Fumes of acetone and concentrated hydrochloric acid are objectionable and, to some degree, harmful. These chemicals should be used in the hood. Be sure to wear safety glasses.

After you have finished the experiment, dispose of the solvent mixture in a stoppered bottle.

PROCEDURE

1. Using a graduated cylinder and working in the hood, prepare the following solvent system: 19 mL acetone; 4 mL concentrated hydrochloric acid, HCl; 2 mL water. Pour the solvent mixture into an 800- or 1000-mL beaker and cover it tightly with a plastic film (e.g. Saran wrap). This allows the atmosphere within the beaker to become saturated with solvent vapor and helps to give a better chromatographic separation.
2. Obtain a piece of chromatography paper 24 to 25 cm long by 11 to 14 cm wide. Draw a pencil line about 1 cm from the long edge of the paper. (You must use an ordinary pencil for this line. Ink or colored pencil often contains substances that may be soluble in the solvent, producing chromatograms of their own.) This line will indicate the origin (see Figure 4–1). Also draw a line about 1 cm long, 6 cm above the center of the penciled line.
3. Using a different capillary tube for each solution (do not mix them!) transfer a drop of each solution listed below to the penciled line, as shown in Figure 4–1. Apply the spots evenly over the line, leaving a margin of about 3 cm from each short edge of the paper. Use a separate, clean capillary tube for each solution; or, if the solutions are to be obtained from beakers in which a capillary tube is provided, *be sure to return the tube to its proper beaker.* With a pencil, identify each spot by writing on the paper directly beneath the spot. The solutions are:
   A) Fe^{3+} solution
   B) Cu^{2+} solution
   C) Co^{2+} solution
   D) Solution containing all three ions, Fe^{3+}, Cu^{2+}, and Co^{2+}
   E) Any of the unknowns furnished (be sure to record its number)
   F) Another unknown (again, record the number)
4. Dry the paper under a heat lamp or air blower.
5. Form the paper into a cylinder without overlapping the edges. Fasten the paper with staples, as shown in Figure 4–2.

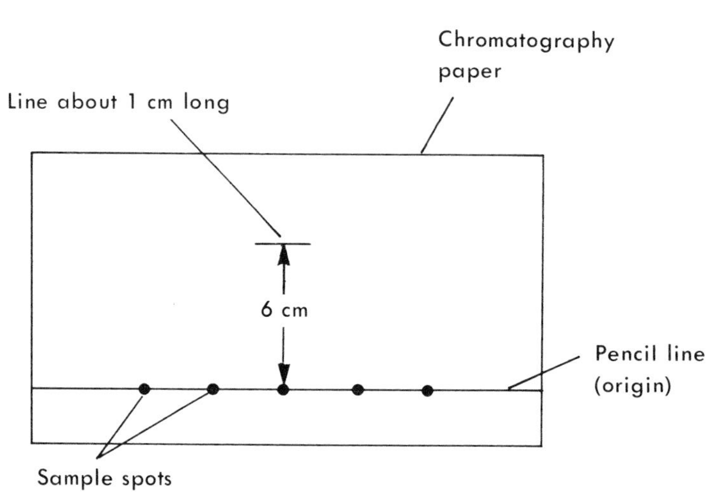

Figure 4–1. Preparing chromatography paper.

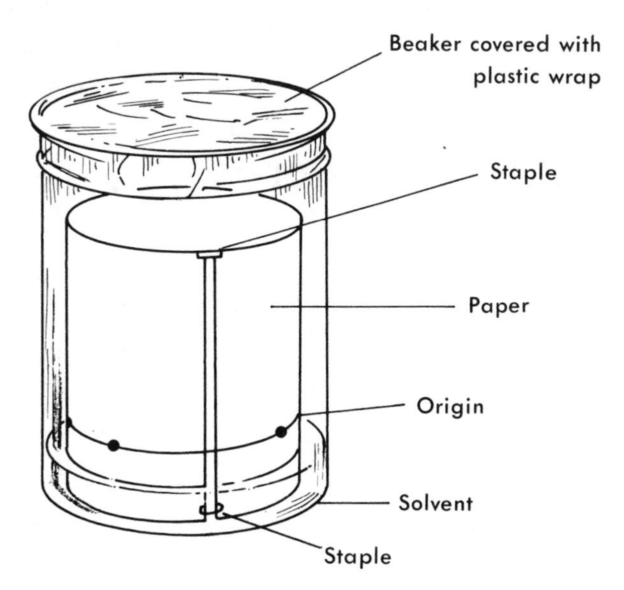

Figure 4–2. Development of chromatogram.

6. Place the beaker in a position on your desk where it can remain undisturbed throughout this step. Taking care that the origin line remains above the solvent, carefully place the cylinder into the beaker, as shown in Figure 4–2. Replace the plastic film and wait as the solvent moves up the paper. Do not move the beaker.
7. *NOTE: In this and all remaining steps, when the paper is wet, be sure not to lay it down on any surface that is not clean.* When the solvent has risen above the short line drawn 6 cm above the origin in Step 2, remove the cylinder from the beaker and quickly mark the solvent front position with a pencil. Remove the staples and dry the paper under a heat lamp.
8. Spray the paper with a solution of potassium hexacyano ferrate(II), $K_4[Fe(CN)_6]$.*

   The presence of Fe^{3+} is shown by the spot turning a dark steel blue color. Cu^{2+} turns a rust brown, and Co^{2+} turns an apple-green color.

*Also called potassium ferrocyanide.

RESULTS AND CALCULATIONS

Observe and record on the report sheet the colors of spots produced by the three ions in each of the chromatograms of solutions A through D.

Measure and record in millimeters the distance between the origin and the solvent front (X in Figure 4–3). Next, measure and record the distance between the origin and the center of each spot in the chromatograms for solutions A through D. Calculate the R_F value for each ion, using Equation 4.1. Record that value as a decimal fraction to the proper number of significant figures.

From the spots above your solutions E and F, indicate by “Yes” or “No” in the table the ions present in the unknowns. Be sure to list the identification numbers of the unknowns.

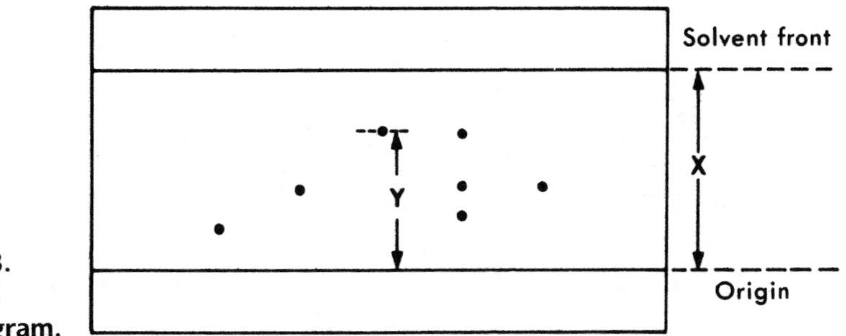

Figure 4–3. Developed chromatogram.

Experiment 4

Work Page

NAME ....

DATE .... SECTION ....

Distance between origin and solvent front: _____________ mm

SOLUTIONS OF KNOWN IONS

Solution	*Ion*	*Color*	*Distance from Origin (mm)*	R_F
A	Fe^{3+}			
B	Cu^{2+}			
C	Co^{2+}			
D	Fe^{3+}			
	Cu^{2+}			
	Co^{2+}			

SOLUTIONS OF UNKNOWN IONS

Enter "Yes" or "No" to indicate the presence or absence of each ion. Be sure to enter the identification number of your unknown.

Solution	*Unknown No.*	Fe^{3+}	Cu^{2+}	Co^{2+}
E				
F				

Experiment 4

Report Sheet

NAME ....

DATE .... SECTION ....

Distance between origin and solvent front: ______________ mm

SOLUTIONS OF KNOWN IONS

Solution	*Ion*	*Color*	*Distance from Origin (mm)*	R_F
A	Fe^{3+}			
B	Cu^{2+}			
C	Co^{2+}			
D	Fe^{3+}			
	Cu^{2+}			
	Co^{2+}			

SOLUTIONS OF UNKNOWN IONS

Enter "Yes" or "No" to indicate the presence or absence of each ion. Be sure to enter the identification number of your unknown.

Solution	*Unknown No.*	Fe^{3+}	Cu^{2+}	Co^{2+}
E				
F				

Experiment 4

Advance Study Assignment

NAME ....

DATE .... SECTION ....

1) What would you observe if you used a ballpoint pen, instead of a pencil, to mark the chromatography paper?

2) Why do you have to cover the beaker while the solvent is moving up the paper?

3) Make the necessary measurements (in millimeters) and calculate R_F for the following chromatogram:

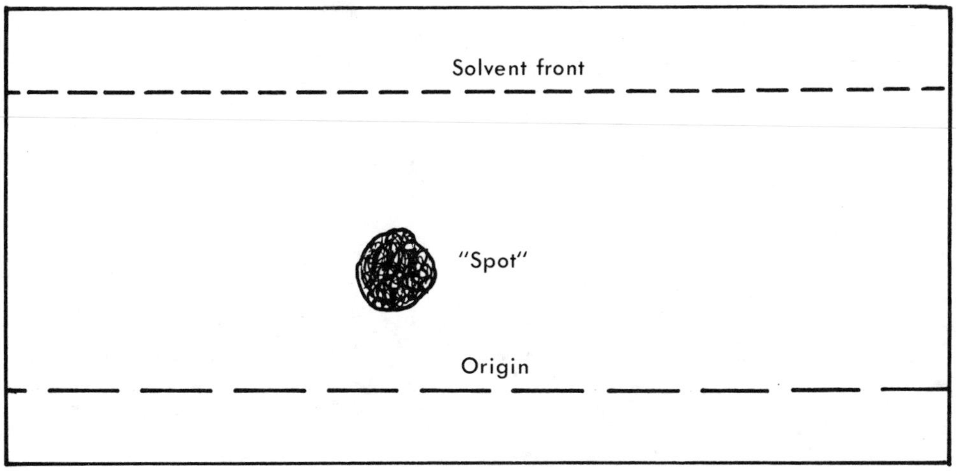

Experiment 5

Densities of Liquids and Solids

Performance Goal

5–1 Calculate the density of a liquid or a solid from experimental data.

CHEMICAL OVERVIEW

One of the physical properties that characterize a substance is its **density,** which is defined as its *mass per unit volume.* Mathematically,

$$\text{Density} = \frac{\text{mass}}{\text{volume}} \tag{5.1}$$

According to this equation, density is equal to the ratio of the mass of a sample of a substance to the volume it occupies. The density of a solid is normally expressed in grams per cubic centimeter (g/cm^3), the density of a liquid in grams per cubic centimeter or grams per milliliter (g/mL), and the density of a gas in grams per liter (g/L).

To determine the density of a substance you must measure both the mass and volume of the same sample of the substance. Density is then calculated by dividing the mass by the volume, as indicated in Equation 5.1. Mass is measured by the usual weighing techniques. The volume of a liquid may be measured in a graduated cylinder. The dimensions of a solid with a regular geometric shape (rectangular block, cylinder, sphere) may be measured with a ruler, and these measurements can then be used to calculate the volume. The volume of a solid with an irregular shape may be determined by measuring the volume of a liquid displaced when the solid is immersed in the liquid.

In Part 1 of this experiment you will be asked to determine experimentally the density of a known substance and then to calculate the **percent error** in your determination. Percent error is defined by the following equation:

$$\text{Percent error} = \frac{\text{error}}{\text{accepted value}} \times 100 \tag{5.2}$$

The "error" is the difference between the experimental value and the accepted value. Error is expressed as an **absolute value,** i.e., a numerical value without regard for algebraic sign. Absolute value is indicated by enclosing the quantity between vertical lines. Thus Equation 5.2 becomes

$$\text{Percent error} = \frac{|\text{experimental value} - \text{accepted value}|}{\text{accepted value}} \times 100 \tag{5.3}$$

SAFETY PRECAUTIONS AND DISPOSAL METHODS

Safety hazards in this experiment cannot be identified precisely because of the wide variety of chemicals that might be used as liquid unknowns. This uncertainty dictates that all liquids be regarded as potentially dangerous and treated accordingly. This includes the known liquid, trichloroethane. Liquid samples should be obtained from a dispensing station in the hood. If taken from the hood, liquids should be in containers that are stoppered or covered with a plastic sheet or metal foil. Some unknowns may be flammable; they should therefore be kept away from open flames. When you are finished with them, discard them as directed by your instructor. Avoid contact between all liquids and your skin; if it occurs, wash the exposed area thoroughly with soap and water. *Safety glasses must be worn at all times.*

Depending on the nature of your liquid unknown, disposal directions will be given by your instructor. Trichloroethane should be collected in a stoppered bottle.

PROCEDURE

> **Note: All mass measurements are to be recorded in grams to the nearest 0.01 g. Length measurements are to be recorded in centimeters to the nearest 0.1 cm. Record liquid volume measurements in milliliters to the nearest 0.1 mL.**

1. DENSITY OF A LIQUID

A) Your 50-mL graduated cylinder and a piece of plastic wrap (e.g. Saran wrap) to cover the opening constitute your "container" (see page 8) for Part 1 of this experiment. Being sure the cylinder is clean and dry, weigh it and the Saran wrap — the container — to the nearest 0.01 g on a centigram balance. Record the mass in the proper number of significant figures on your report sheet.

B) Take the cylinder and plastic wrap to the hood. Pour 12 to 15 mL of 1,1,1-trichloroethane into the cylinder; do *not* attempt to make the amount *exactly* 12, 13, 14, or 15 mL. Cover the cylinder with the plastic wrap. Estimate the volume to the nearest 0.1 mL (see page 7 on reading volume), and record that value to the proper number of significant figures.

C) After making sure the outside of the cylinder is dry, measure and record the mass of the container plus liquid on the centigram balance.

D) Dispose of your liquid as directed by your instructor.

E) In the same manner, collect data for finding the densities of one or more unknown liquids, as required by your instructor. *Be sure to record the identification number of each unknown.*

2. DENSITY OF A REGULAR SOLID

Select one or more of the solid unknowns provided for this experiment and record its identification number. Determine and record its mass to the nearest centigram. Make whatever measurements may be necessary to calculate the volume of the object, listing these measurements to the closest 0.1 cm. Because these objects are of various shapes, the data table contains blank spaces in which to describe the shapes and identify the measurements (length, diameter, etc.) that are made.

3. DENSITY OF AN IRREGULAR SOLID

A) Place 20 to 25 mL of water into the cylinder from Part 1. Record the volume to the nearest 0.1 mL. Determine the mass of the cylinder plus water to the nearest centigram. This is the mass of the container for Part 3.

B) Select and record the identification number of one of the unknown irregular solids provided for this experiment. Place enough of the solid into the graduated cylinder to cause the liquid level to rise by more than 10 milliliter markings. Be sure all of the solid is below the surface of the liquid. Record the volume to the nearest 0.1 mL. Also measure the mass of the container and its contents to the nearest centigram.

C) Dispose of your solid material into the recovery facility that has been set up in your laboratory. Be careful not to mix unknown solids.

D) Repeat Steps 3A and 3B for as many unknowns as are required by your instructor, or for a second run with the same unknown.

CALCULATIONS

Be sure to include units in the results of all calculations. Also be sure to express those results in the correct number of significant figures.

1. DENSITY OF A LIQUID

Find the mass of the liquid by difference — by subtracting the mass of the container from the mass of the container plus liquid. The density of the liquid is found by dividing the mass of the liquid sample by its volume, as indicated in Equation 5.1. Percent error may be calculated by substituting into Equation 5.3; be careful of significant figures in the result. The accepted value for the density of 1,1,1-trichloroethane is 1.34 g/mL.

2. DENSITY OF A REGULAR SOLID

The volume of a rectangular solid is calculated by multiplying the length by the width by the height: $V = l \times w \times h$.

The volume of a cylinder is the area of the base times the height. The area of a circle is $\pi d^2/4$, where d is the diameter. Thus

$$V_{cylinder} = \frac{\pi d^2 h}{4}$$

The value of π to eight decimal places is 3.14159265; the number of places to which you should round it off is left to you.

The volume of a sphere is found from the following equation:

$$V_{sphere} = \frac{\pi d^3}{6}$$

Once you have calculated the volume of the unknown solid, its density may be found by substituting into Equation 5.1, as before.

3. DENSITY OF AN IRREGULAR SOLID

Both the mass and the volume of the sample are found by difference. Density is again calculated by substitution into Equation 5.1.

Experiment 5
Work Page

NAME ....

DATE .... SECTION ....

1) DENSITY OF A LIQUID

Liquid (List identification number of unknowns)	***Trichlorethane***			
Mass of container + liquid (g)				
Mass of container (g)				
Mass of liquid (g)				
Volume of liquid (mL)				
Density (g/mL)				
Percent error (trichloroethane only)				

"Accepted" value for density of 1,1,1-trichloroethane: 1.34 g/mL.

Calculation Setups for Density Determinations:

Calculation Setups for Percent Error for Trichloroethane:

2) DENSITY OF A REGULAR SOLID

Unknown Number				
Shape of unknown				
Volume of unknown (cm^3)				
Mass of unknown (g)				
Density (g/cm^3)				

Calculation Setups for Determination of Volumes of Unknowns:

For each unknown, list the measurements taken and show calculation setup.

Calculation Setups for Density Determinations:

Experiment 5

Work Page

NAME ....

DATE .... SECTION ....

3) DENSITY OF AN IRREGULAR SOLID

Unknown Number				
Mass of container + liquid + solid (g)				
Mass of container + liquid (g)				
Mass of solid (g)				
Volume of liquid + solid (mL)				
Volume of liquid (mL)				
Volume of solid (cm^3)				
Density (g/cm^3)				

Calculation Setups for Density Determinations:

Experiment 5

Report Sheet

NAME ....

DATE .... SECTION ....

1) DENSITY OF A LIQUID

Liquid (List identification number of unknowns)	*Trichlorethane*			
Mass of container + liquid (g)				
Mass of container (g)				
Mass of liquid (g)				
Volume of liquid (mL)				
Density (g/mL)				
Percent error (trichloroethane only)				

"Accepted" value for density of 1,1,1-trichloroethane: 1.34 g/mL.

Calculation Setups for Density Determinations:

Calculation Setups for Percent Error for Trichloroethane:

2) DENSITY OF A REGULAR SOLID

Unknown Number				
Shape of unknown				
Volume of unknown (cm^3)				
Mass of unknown (g)				
Density (g/cm^3)				

Calculation Setups for Determination of Volumes of Unknowns:

For each unknown, list the measurements taken and show calculation setup.

Calculation Setups for Density Determinations:

Experiment 5

Report Sheet

NAME ....

DATE .... SECTION ....

3) DENSITY OF AN IRREGULAR SOLID

Unknown Number				
Mass of container + liquid + solid (g)				
Mass of container + liquid (g)				
Mass of solid (g)				
Volume of liquid + solid (mL)				
Volume of liquid (mL)				
Volume of solid (cm^3)				
Density (g/cm^3)				

Calculation Setups for Density Determinations:

NAME ....

Experiment 5

Advance Study Assignment

DATE .... SECTION ....

1) The volume of an unknown liquid is 28.6 mL and its mass is 32.2 grams. Calculate the density of the liquid.

2) When 95.0 g of an unknown metal are submerged in water in a graduated cylinder, the water level rises from 38.2 mL to 49.5 mL. Calculate the density of the metal.

3) The accepted value for the density of a certain metal is 5.48 g/cm^3. Calculate the percent error in a laboratory experiment that yields a value of 5.2 g/cm^3. Express this result in the proper number of significant figures.

Experiment 6Ⓜ

Simplest Formula of a Compound

Performance Goals

6–1 Prepare a compound and collect data from which you can determine the mass of each element in the compound.
6–2 From the mass of each element in a compound, determine its simplest formula.

CHEMICAL OVERVIEW

Chemical compounds are composed of atoms of different elements. The atoms are held together by chemical bonds. It has been shown experimentally that the ratio of moles of the elements in a compound is nearly always a ratio of small, whole numbers. The few exceptions are known as nonstoichiometric compounds. The formula containing the lowest possible ratio is known as its **simplest formula.** It is also called the **empirical formula.** At times it may be the same as the molecular formula; often, however, the molecular formula is an integral multiple of the simplest, empirical formula. For example, the simplest formula of the compound benzene (C_6H_6) is simply CH, indicating that the ratio of carbon atoms to hydrogen atoms is one to one.

To find the simplest formula of a compound, you will combine the elements in the compound under conditions that will allow you to determine the mass of each element. From these data the moles of atoms of each element may be calculated. By dividing these numbers by the smallest number of moles, you obtain quotients that are in a simple ratio of integers, or are readily converted to such a ratio. The ratio of moles of atoms of the elements in a compound is the same as the ratio of individual atoms that is expressed in the empirical formula.

Remember: The essential information you require to find the simplest formula of a compound is the number of grams of each element in a sample of the compound.

In Option 1 you will react a measured mass of copper with excess sulfur. The excess sulfur is burned away as sulfur dioxide. In Option 2 the reaction is between a measured quantity of tin and excess nitric acid. The excess acid is boiled off. Option 3 involves the reaction of a measured mass of magnesium with excess oxygen from the air.

Your instructor may require you to perform the experiment twice to obtain duplicate results, or to complete more than one option. If so, plan your use of time. The procedure includes some periods in which you wait for a crucible to cool. The cooling periods in the first run of the experiment can be used for heating periods in the second run, and vice versa.

SAMPLE CALCULATIONS

A piece of aluminum is ignited in a suitable container, yielding an oxide. Calculate the simplest formula of the oxide from the following data:

Mass of container	17.84 g
Mass of container + aluminum	18.38 g
Mass of container + compound	18.86 g

1) Mass of each element from data:

$$\text{Mass of container + aluminum} - \text{mass of container} = \text{mass of aluminum}$$
$$18.38\ \text{g} - 17.84\ \text{g} = 0.54\ \text{g aluminum}$$

$$\text{Mass of container + compound} - \text{mass of container + aluminum} = \text{mass of oxygen}$$
$$18.86\ \text{g} - 18.38\ \text{g} = 0.48\ \text{g oxygen}$$

2) Moles of each element:

$$0.54\ \text{g Al} \times \frac{1\ \text{mole Al atoms}}{27.0\ \text{g Al}} = 0.020\ \text{mole Al atoms}$$

$$0.48\ \text{g O} \times \frac{1\ \text{mole O atoms}}{16.0\ \text{g O}} = 0.030\ \text{mole O atoms}$$

3) Simplest formula ratio:

Obtain the ratio of atoms by dividing the number of moles of each atom by the smallest number of moles:

$$\text{Al:}\ \frac{0.020}{0.020} = 1.0; \qquad \text{O:}\ \frac{0.030}{0.020} = 1.5$$

The ratio is 1.0 mole aluminum atoms to 1.5 moles oxygen atoms. Change this ratio to a whole number ratio by multiplying each value by 2:

$$\text{moles of Al atoms} = 1.0 \times 2 = 2.0$$

$$\text{moles of O atoms} = 1.5 \times 2 = 3.0$$

The simplest formula is therefore Al_2O_3.

Many students find it convenient to organize their calculations by arranging both data and results in a table as follows:

Element	*Grams*	*Moles*	*Mole Ratio*	*Formula Ratio*
Al	0.54	0.020	1.0	2
O	0.48	0.030	1.5	3

SAFETY PRECAUTIONS AND DISPOSAL METHODS

The safety considerations in this experiment relate to the operation of a Bunsen burner and the handling of hot items. Blue burner flames are visible, but easily lost against a laboratory background. Be careful not to reach through one in reaching for some object behind it. If you have long hair, tie it back so it does not get into the flame. Be sure to use crucible tongs in handling hot crucibles, including the lid. Laboratory hardware gets hot, too. Harmful gases are released in Options 1 and 2. These reactions must be performed in a fume hood, as stated in their procedures. Be careful of hot chemical spattering from crucibles when they are heated. Be sure to wear goggles throughout this experiment.

Dispose of any solid residue as directed by your instructor.

PROCEDURE

OPTION 1: A SULFIDE OF COPPER

Note: All mass measurements in Option 1 are to be recorded in grams to the nearest 0.01 g.

A) The purpose of this step is to remove moisture from the crucible. Support a clean, dry porcelain crucible and its lid on a clay triangle, as shown in Figure 6–1. Heat it slowly at first, and then fairly strongly in the direct flame of a burner for about 4 to 5 minutes. Set the crucible and lid aside on a wire gauze to cool.

B) When the crucible and lid are cool to the touch, weigh them on a centigram balance. Record this value as the mass of the container.

C) Place a loosely rolled ball of copper wire or medium shavings, about 1.5 to 2 g, into the crucible. Weigh them, with the lid, on a centigram balance, and record the weight as the mass of the container plus metal.

D) Sprinkle about 1 to 1.5 g of powdered sulfur over the copper. Place the lid on the crucible and begin heating it in a fume hood. Heat slowly at first, and then with a moderate flame until the sulfur no longer burns around the lid. Finally, heat the crucible strongly for about 5 minutes, making sure that no excess sulfur is present on the lid or on the sides of the crucible. It sometimes helps to hold the burner at its base and direct the flame under the lip of the lid all the way around.

E) Set the container and its contents aside to cool. Do not open the lid until the crucible is cool, as air oxidation is apt to occur.

F) When the crucible is cool, lift the lid and examine the contents. There should be no evidence of sulfur in the crucible or on the lid. If sulfur is present, heat the crucible again until the sulfur is completely burned away. Allow the crucible to cool.

G) Weigh the container and its contents again. Record your measurement as the mass of the container plus compound.

H) Set the container aside while you complete your calculations. Do not discard your compound until your calculations are finished and satisfactory; if they are not satisfactory, it is possible you may be able to salvage your work if the material is still on hand.

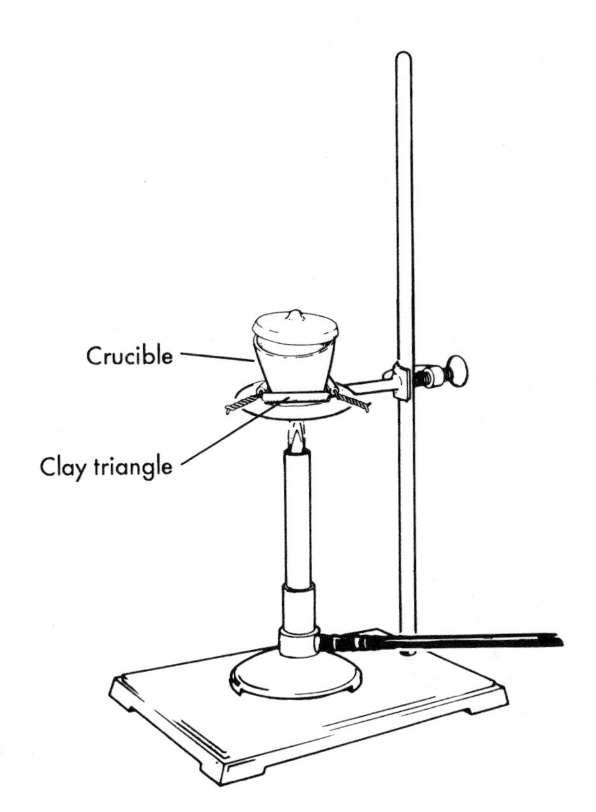

Figure 6–1. Heating of a porcelain crucible.

I) Just before discarding the compound, press it to the bottom of the crucible. Notice the difference between the physical properties of the compound and those of the elements from which it was formed. The compound should be discarded as directed by your instructor.

OPTION 2: AN OXIDE OF TIN

Note: All mass measurements in Option 2 are to be recorded in grams to the nearest 0.01 g.

A) Heat a porcelain crucible and its lid as described in Option 1, Step A. Allow it to cool.

B) Weigh as described in Option 1, Step B.

C) Place a loosely rolled ball of tin foil, weighing 1 to 1.5 g, into the crucible. Weigh the crucible, the lid, and the metal on a centigram balance. Record the mass obtained.

D) Under the fume hood, add concentrated nitric acid, HNO_3, drop by drop, to the crucible until all the tin has reacted and a damp white paste remains.

E) Heat the paste cautiously with a mild flame, taking care not to cause spattering. After all of the liquid has evaporated, heat the crucible with a hot flame for 5 minutes.

F) Cool the crucible and compound to room temperature and weigh it. Record the mass of the container and compound.

G) Keep your compound in the crucible until all calculations are completed. This may save you time if it becomes necessary to add more nitric acid.

OPTION 3: MAGNESIUM OXIDE

Note: All mass measurements in Option 3 are to be recorded in grams to the nearest 0.01 g.

A) Heat a porcelain crucible and its lid as described in Option 1, Step A. Allow it to cool.

B) Weigh the crucible and lid on a *milligram* balance. Record this value on the report sheet as mass of the container.

C) Place a loosely folded magnesium ribbon, weighing 0.5 to 0.7 g, into the crucible. Weigh the crucible, lid, and metal on a milligram balance and record the mass.

D) Remove the lid and hold it near the crucible with a pair of tongs. Start heating the crucible, and as soon as the magnesium begins to burn, replace the lid. Continue the process, holding the escape of white smoke to a minimum (very finely divided magnesium oxide looks like smoke). When the contents of the crucible no longer burn, cock the lid wide enough to allow a sufficient amount of air to enter to complete the reaction, as shown in Figure 6–1, and heat it strongly for 5 minutes.

E) To convert the possible side product, magnesium nitride, to the oxide, let the crucible cool, add 10 drops of deionized water to it and then gently heat to vaporize excess water. CAUTION: SPATTERING MAY OCCUR.

F) Finish heating the crucible with a strong flame for 5 to 8 minutes.

G) Weigh the cool crucible, lid, and product on a milligram balance.

Experiment 6
Work Page

NAME ....

DATE .... SECTION ....

DATA

Option or Trial				
Mass of container (g)				
Mass of container + metal (g)				
Mass of container + compound (g)				

RESULTS FOR SAMPLE OF COMPOUND PREPARED

Mass of metal (g)				
Mass of nonmetal element (g)				
Moles of metal				
Moles of nonmetal element				
Ratio: $\frac{\text{moles metal}}{\text{moles nonmetal}}$ *	$\frac{\quad}{1}$	$\frac{\quad}{1}$	$\frac{\quad}{1}$	$\frac{\quad}{1}$
Simplest formula				

* Express this ratio as a decimal number over 1 (e.g., $\frac{3.044}{1}$), with the numerator to the number of significant figures justified by the data.

Show calculations on the reverse side of this page.

CALCULATIONS

Experiment 6

Report Sheet

NAME ....

DATE .... SECTION ....

DATA

Option or Trial				
Mass of container (g)				
Mass of container + metal (g)				
Mass of container + compound (g)				

RESULTS FOR SAMPLE OF COMPOUND PREPARED

Mass of metal (g)				
Mass of nonmetal element (g)				
Moles of metal				
Moles of nonmetal element				
Ratio: $\frac{\text{moles metal}}{\text{moles nonmetal}}$ *	$\frac{\quad}{1}$	$\frac{\quad}{1}$	$\frac{\quad}{1}$	$\frac{\quad}{1}$
Simplest formula				

* Express this ratio as a decimal number over 1 (e.g., $\frac{3.044}{1}$), with the numerator to the number of significant figures justified by the data.

Show calculations on the reverse side of this page.

CALCULATIONS

NAME ....

Experiment 6

Advance Study Assignment

DATE .... SECTION ....

1) Circle the one of the following formulas that is correctly written as an empirical formula:

$NaSO_{1.5}$ $(H_2NO)_2$ Fe_3O_4

2) 6.25 grams of pure iron are allowed to react with oxygen to form an oxide. If the product weighs 14.31 grams, find the simplest formula of the compound.

3) In determining the simplest formula of lead sulfide, 2.46 grams of lead are placed in a crucible with 2.00 grams of sulfur. When the reaction is complete, the product has a mass of 3.22 grams. What mass of sulfur should be used in the simplest formula calculation? Find the simplest formula of lead sulfide.

Experiment 7Ⓜ

Hydrates

Performance Goals

7–1 Calculate the percentage of water of hydration in a compound from experimental data.
7–2 Calculate the formula of a hydrate of a known anhydrous salt from experimental data.

CHEMICAL OVERVIEW

Hydrates are chemical compounds that contain water as part of their crystal structure. This water is quite strongly bound, and is present in a definite proportion relative to other constituents. It is referred to as **water of hydration.**

The formula of a hydrate consists of the formula of the **anhydrous** (without water) **compound** followed by a dot, the number of water molecules that crystallize with one formula unit of the compound, and the formula of water. For example, $CuSO_4 \cdot 5H_2O$ indicates that 5 molecules of water — H_2O — crystallize with 1 formula unit of anhydrous copper(II) sulfate, $CuSO_4$, to form copper sulfate pentahydrate, or, by its alternative names, copper sulfate 5 water or copper sulfate 5 hydrate.

Generally, water of hydration can be driven from hydrates by heating, leaving behind the anhydrous salt. The process may be accompanied by physical changes, such as a change in color. For example, $CuSO_4 \cdot 5H_2O$ is an intense blue shiny crystal which, upon heating, turns into pale green-blue anhydrous $CuSO_4$.

In this experiment you will be instructed to determine the mass of a sample of an unknown hydrate by difference, using a preweighed crucible as the container. The substance will be "dehydrated" by heat, and weighed again. The loss of mass represents the mass of water in the original sample, which may be expressed as percentage of water of hydration, using the usual (part quantity/total quantity) × 100 relationship:

$$\begin{array}{c}\text{Percentage water}\\ \text{of hydration}\end{array} = \frac{\text{grams of water}}{\text{grams of hydrate}} \times 100 \qquad (7.1)$$

To find the formula of the original hydrate, you will determine from the preceding data the mass of the anhydrous compound, the formula of which will be given to you. From this you can calculate the moles of anhydrous compound in the original sample. From the mass of water in the original sample you can calculate the moles of water. By dividing the moles of water by the moles of anhydrous salt, you obtain the ratio of moles of water to 1 mole of anhydrous salt, as in the 5-to-1 ratio for $CuSO_4 \cdot 5H_2O$ above.

In this experiment you will be directed to "heat to constant mass." Your purpose is to heat the substance until *all* of the water is driven off. After a first heating, cooling, and weighing, you cannot tell if all water has been removed, or if some still remains. You therefore repeat the heating, cooling, and weighing procedure. If the same mass is reached after the second heating, you may assume that all water was removed the first time. If mass was lost in the second heating, you may be sure that all water was *not* removed in the first heating, and you are still unsure that it was all driven off in the second heating. Another heating is therefore required. The heating, cooling, and weighing sequence is repeated until two successive duplicate weighings are recorded. Two weighings within the ± uncertainty range of the balance are generally considered to be duplicate weighings; duplication within 0.005 g is satisfactory for this experiment.

Your instructor may require you to perform the experiment twice to obtain duplicate results, or to run more than one unknown. If so, plan your use of time. The procedure includes some periods in which you must wait for a crucible to cool. The cooling periods in the first run of the experiment can be used for heating periods in the second run, and vice versa. In this way you perform both runs simultaneously.

SAFETY PRECAUTIONS AND DISPOSAL METHODS

The safety considerations in this experiment relate to the operation of a Bunsen burner and the handling of hot items. Blue burner flames are visible, but easily lost against a laboratory background. Be careful not to reach through a flame in picking up some object behind it. If you have long hair, tie it back so it does not get into the flame. Be sure to use crucible tongs in handling hot crucibles, including the lids. Laboratory hardware gets hot, too. Goggles should always be worn when working with chemicals, and particularly while heating them, as in this experiment. Be alert also to the possibility of hot chemicals "shooting" out of the test tube in Part 2.

The anhydrous solids should be discarded in a container or you should follow directions given by your instructor.

PROCEDURE

Note: Record all mass measurements in grams to the nearest 0.001 g.

1. PERCENTAGE WATER IN A HYDRATE: FORMULA OF A HYDRATE

A) Heat a clean porcelain crucible and its lid — the "container" for this experiment (see pages 8 and 9) — on a clay triangle over a direct flame for 5 minutes to drive off any surface moisture. (See Figure 6–1, page 75.) When they are cool to the touch, weigh them on a milligram balance, recording this and subsequent masses in the "constant mass" portion of your data sheet. Heat the container to constant mass in 5-minute heating cycles until duplicate masses (within 0.005 g) are reached. Record the final (constant) mass as the mass of the container.

Note: Part 2 of the experiment may be performed during the cooling cycles of Part 1.

B) Place 1 to 1.5 g of the unknown solid hydrate into the container, and weigh again on a milligram balance. Record as the mass of the container plus hydrate.

C) With the lid almost covering the crucible, heat the container and its contents, gently at first, and then with a hot flame for 10 minutes. When they feel cool to the touch, weigh them again. Record this and subsequent weighings in the "constant mass" portion of the data sheet. Heat to constant mass in 4-to-6-minute heating cycles until duplicate masses are reached. The final (constant) mass should be recorded as the mass of the container plus anhydrous salt.

D) Set the container and its contents aside while you complete your calculations. Do not discard the residue until your calculations are finished and satisfactory; if they are not satisfactory, it is possible that you may be able to salvage your work if the material is still on hand.

2. BEHAVIOR OF A HYDRATE

A) Place a few small crystals of $CoCl_2 \cdot 6H_2O$ into a test tube. Holding the test tube tilted at an angle, and with its mouth pointing away from you and all others, heat the test tube gently. Record your observations.

B) After the test tube has cooled to room temperature, add a few drops of water. Hold the test tube against the back of your hand. Record your observations.

CALCULATIONS

From the masses of the container, the container plus hydrate, and the container plus anhydrous salt, identify and perform the subtractions that will yield the mass of the hydrate and the mass of the water of hydration. From these, calculate the percentage water of hydration, using Equation 7.1.

Return to the original data (masses of container, container plus hydrate, and container plus anhydrous salt), and identify and perform the subtraction that will yield the mass of the anhydrous salt. The anhydrous salts resulting from dehydration of the different unknowns in this experiment are:

UNKNOWN	yields	ANHYDROUS SALT
A		$BaCl_2$
B		$MgSO_4$
C		$CuSO_4$

Convert the grams of anhydrous salt to moles. Also convert grams of water of hydration to moles. To the number of significant figures justified by your data, calculate the number of moles of water per mole of anhydrous compound; enter this result in the "Ratio" line in your table of results. From the ratio, determine the formula of the hydrate.

Experiment 7
Work Page

NAME ....

DATE .... SECTION ....

Data and result tables for Experiment 7
appear on pages 88 and 89

1) PERCENTAGE WATER IN A HYDRATE; FORMULA OF A HYDRATE

Constant Mass Data (Supply identification letters or numbers for unknowns)

Item	*Container*	*Unknown* ______	*Unknown* ______	*Unknown* ______	*Unknown* ______
1st heating (g)					
2nd heating (g)					
3rd heating (g)					
4th heating (g)					
Final heating (g)					

Mass Data

Container (g)				
Container + hydrate (g)				
Container + anhydrous salt (g)				

Results (Show full calculations for one column at the top of the next page)

Mass of hydrate (g)				
Mass of water of hydration (g)				
Percentage water of hydration				
Mass of anhydrous salt (g)				
Formula of anhydrous salt				
Moles of anhydrous salt				
Moles of water of hydration				
Ratio: $\frac{\text{moles of water}}{\text{1 mole of anhydrous salt}}$ *	$\frac{\_\_}{1}$	$\frac{\_\_}{1}$	$\frac{\_\_}{1}$	$\frac{\_\_}{1}$
Formula of hydrate				

* Express this ratio as a decimal number over 1 (e.g., $\frac{2.875}{1}$) with the numerator to the number of significant figures justified by the data.

Experiment 7
Work Page

NAME ....

DATE .... SECTION ....

Calculations for One Full Column of Results from Previous Page

2) BEHAVIOR OF A HYDRATE

A) Observations when a hydrate is heated in a test tube.

B) Observations when water is added to an anhydrous salt.

Experiment 7

Report Sheet

NAME ....

DATE .... SECTION ....

Data and result tables for Experiment 7
appear on pages 92 and 93

1) PERCENTAGE WATER IN A HYDRATE; FORMULA OF A HYDRATE

Constant Mass Data (Supply identification letters or numbers for unknowns)

Item	*Container*	*Unknown* ______	*Unknown* ______	*Unknown* ______	*Unknown* ______
1st heating (g)					
2nd heating (g)					
3rd heating (g)					
4th heating (g)					
Final heating (g)					

Mass Data

Container (g)				
Container + hydrate (g)				
Container + anhydrous salt (g)				

Results (Show full calculations for one column at the top of the next page)

Mass of hydrate (g)				
Mass of water of hydration (g)				
Percentage water of hydration				
Mass of anhydrous salt (g)				
Formula of anhydrous salt				
Moles of anhydrous salt				
Moles of water of hydration				
Ratio: $\frac{\text{moles of water}}{\text{1 mole of anhydrous salt}}$ *	$\frac{\quad}{1}$	$\frac{\quad}{1}$	$\frac{\quad}{1}$	$\frac{\quad}{1}$
Formula of hydrate				

* Express this ratio as a decimal number over 1 (e.g., $\frac{2.875}{1}$) with the numerator to the number of significant figures justified by the data.

Experiment 7
Report Sheet

NAME ....

DATE .... SECTION ....

Calculations for One Full Column of Results from Previous Page

2) BEHAVIOR OF A HYDRATE

A) Observations when a hydrate is heated in a test tube.

B) Observations when water is added to an anhydrous salt.

Experiment 7

Advance Study Assignment

NAME ....

DATE .... SECTION ....

1) How can you make sure that all of the water of hydration has been removed? Explain.

2) A 2.815-g sample of $CaSO_4 \cdot XH_2O$ was heated until all of the water was removed. Calculate the percentage water of hydration and the formula of the hydrate if the residue after heating weighed 2.485 g.

Experiment 8

Percentage of Oxygen in Potassium Chlorate

Performance Goal

8–1 Determine the percentage of one part of a compound from experimental data.

CHEMICAL OVERVIEW

The thermal decomposition of potassium chlorate is described by the equation

$$2\ KClO_3(s) \xrightarrow{\Delta} 2\ KCl(s) + 3\ O_2\ (g) \tag{8.1}$$

In this experiment you will determine the percentage of oxygen in potassium chlorate. You will compare your experimental result with the theoretical percentage calculated from the formula $KClO_3$.

While potassium chlorate decomposes simply by heating, the reaction is intolerably slow. A catalyst, manganese dioxide, MnO_2, is therefore added to speed the reaction. Although it contains oxygen, the catalyst experiences no permanent change during the reaction and does not contribute measurably to the amount of oxygen generated. As with all catalysts, the quantity present at the end of the reaction is the same as the quantity at the beginning.

The experimental procedure is to weigh a quantity of potassium chlorate, heat it to drive off the oxygen, and then weigh the residue, which is assumed to be potassium chloride. The loss in mass represents the oxygen content of the original potassium chlorate.

The above procedure will be carried out in a test tube. The "container" (see page 8) for this experiment will be more than just the test tube, however. It will include the constant mass of catalyst that remains in the test tube throughout the experiment, plus whatever device is used to hold the test tube and its contents while they are being weighed. If your milligram balance has a suspended pan and there is provision for hanging an object to be weighed, you can clamp the test tube in a test-tube holder and hang the entire assembly from the hook provided. In this case the test-tube holder is a part of the container. If the pan on your milligram balance is supported only from beneath, you can stand your test tube in a small beaker each time it is weighed, and include the beaker in the mass of the container. You must be sure, of course, to use the same beaker for each weighing.

In a thermal decomposition such as this, the product must be "heated to constant mass" before you can be sure the decomposition is complete. After the first heating, cooling, and weighing of the decomposed product, the test tube is heated, cooled, and weighed again. If the two weighings are the same, within the plus-and-minus uncertainty of the equipment used, it may be assumed that all of the oxygen was removed in the first heating. If mass is lost in the second heating, it means that some oxygen remained after the first heating and was driven off in the second. It is possible that some oxygen may still be present after the second heating, too. The procedure is therefore repeated again, as many times as necessary, until there is negligible change in mass (no more than 0.005 g for this experiment) between two consecutive weighings.

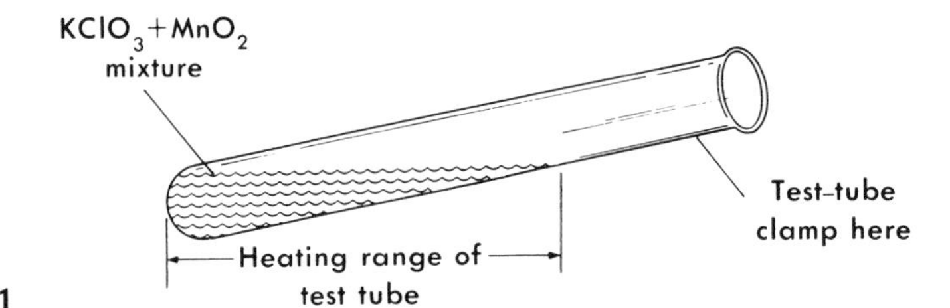

Figure 8–1

SAFETY PRECAUTIONS AND DISPOSAL METHODS

This experiment is potentially hazardous, and if performed carelessly could lead to a serious accident! The formation of a gas at the bottom of a test tube may result in a sudden expansion, blowing hot chemicals out of the test tube. This will not occur if the test tube is handled properly during heating. When heating a solid in a test tube, tip the tube until it is almost horizontal and tap it carefully until the contents are distributed over the lower half of the length of the tube, as shown in Figure 8–1. Holding it at about this angle, move the test tube back and forth in the flame of the burner, distributing the heat over the entire length of the mixture. *Do not concentrate the heat in any one area,* ***particularly near the bottom of the test tube.***

Be very sure your test tube is not pointing toward anybody, including yourself, while it is being heated. Be aware of what those around you are doing while this experiment is being performed in the laboratory. Do not place yourself in the firing line of somebody else's test tube, and if another student points a test tube toward your work station, ask him or her to point it elsewhere. Obviously, *wearing goggles is absolutely mandatory while you or anyone near you is performing this experiment.*

Dispose of the residue in the designated container.

PROCEDURE

Note: Record all mass measurements in grams to the nearest 0.001 g.

A) Place 0.5 to 0.8 g of manganese dioxide, MnO_2, into a large-size test tube. (Do not use a *small* test tube; they are more inclined to "shoot" their contents.) Heat the test tube over a Bunsen burner for about 3 to 4 minutes to drive off any moisture that may be present in the catalyst and test tube. When the tube is cool to the touch, measure the mass of the entire container on a milligram balance. Record the mass in the space provided.

B) Add about 1.0 to 1.5 g of potassium chlorate, $KClO_3$, to the test tube. Find and record the mass of the container and its contents.

C) Mix the contents of the test tube until they have a somewhat uniform gray appearance. (Be careful not to lose any of the contents.) Carefully observing the procedure and safety precaution above, heat the test tube and its contents. Heat gently at first, increasing the intensity after the mixture seems to "boil," as it sometimes appears to do when bubbles of oxygen are being released. Continue heating for about five minutes, and then cool and weigh. Repeat the process in 3- to 5-minute heating cycles until constant mass is reached.

D) Set the container and its contents aside while you complete your calculations. Do not discard the residue until your calculations are finished and satisfactory; if they are not satisfactory, it is possible that you may be able to salvage your work if the material is still on hand.

CALCULATIONS

Using only numbers in the data portion of your report sheet, calculate by difference the initial mass of potassium chlorate and the mass of oxygen released in heating. From these quantities, find the experimental percentage of oxygen in potassium chlorate. Calculate the theoretical oxygen percentage *from the formula of the compound.* Using the theoretical percentage as the accepted value, calculate your percentage error by the equation

$$\text{Percentage error} = \frac{\text{error}}{\text{accepted value}} \times 100$$

$$= \frac{|\text{experimental value} - \text{accepted value}|}{\text{accepted value}} \times 100$$

Note that the numerator of this equation is an absolute value, simply the difference between the observed and theoretical values expressed as a positive quantity.

Experiment 8
Work Page

NAME ....

DATE .... SECTION ....

CONSTANT MASS DATA

Trial	*1*	*2*	*3*	*4*
1st heating (g)				
2nd heating (g)				
3rd heating (g)				
4th heating (g)				
Final heating (g)				

MASS DATA

Mass of container (g)				
Mass of container + $KClO_3$ (g)				
Mass of container + KCl (g)				

RESULTS (Show all calculations for one column on the next page)

Mass of $KClO_3$ used (g)				
Mass of oxygen released (g)				
Percentage oxygen				
Percentage error				

Complete Calculations from One Column of Table:

Calculation of Percent Oxygen in $KClO_3$ from Formula:

Experiment 8

Report Sheet

NAME ....

DATE .... SECTION ....

CONSTANT MASS DATA

Trial	*1*	*2*	*3*	*4*
1st heating (g)				
2nd heating (g)				
3rd heating (g)				
4th heating (g)				
Final heating (g)				

MASS DATA

Mass of container (g)				
Mass of container + $KClO_3$ (g)				
Mass of container + KCl (g)				

RESULTS (Show all calculations for one column on the next page)

Mass of $KClO_3$ used (g)				
Mass of oxygen released (g)				
Percentage oxygen				
Percentage error				

Complete Calculations from One Column of Table:

Calculation of Percent Oxygen in $KClO_3$ from Formula:

NAME ....

Experiment 8

Advance Study Assignment

DATE .... SECTION ....

1) What potential source of laboratory accident is present in this experiment? Explain the procedure you will follow to minimize this hazard. What two precautions will you follow to prevent injury to yourself, and what will you do to avoid injuring another in the event the accident does occur?

2) Another thermal decomposition that produces oxygen begins with silver oxide: $2\ Ag_2O \rightarrow 4\ Ag + O_2$. In a hypothetical experiment, a student collects the following data:

Mass of crucible	30.296 g
Mass of crucible + Ag_2O	38.623 g
Mass of crucible + contents after "complete" decomposition	38.061g

Calculate the following:

A) The starting mass of Ag_2O:

B) The mass of oxygen released:

C) Experimental percentage oxygen in Ag_2O *from data*:

D) Theoretical percentage oxygen in Ag_2O *from the formula*:

E) Percentage error:

Experiment 9

Calorimetry

Performance Goals

9–1 Calculate the specific heat of an unknown solid element by measuring the heat exchanged in a calorimeter.

9–2 Using the Law of Dulong and Petit, calculate the approximate atomic mass of an unknown solid element.

CHEMICAL OVERVIEW

Heat, a form of energy, can be gained or lost by an object. When the object cools, it loses heat energy; when it is heated, it gains energy. The unit in which heat is measured is the **joule, J.** A joule is a derived unit, having base units of $kg \cdot m^2/sec^2$. The joule is a very small amount of heat, so the **kilojoule, kJ,** is commonly used.*

Heat flow is the change in "heat content" of an object as heat energy passes between the object and its surroundings. It is proportional to the mass of the object and its change in temperature. The proportional relationship becomes an equation if the proportionality constant called **specific heat** is introduced:

$$Q = (\text{mass})(\text{specific heat})(\Delta T)$$
$$Q = m \times c \times \Delta T \qquad (9.1)$$

where Q is the heat flow in joules, m is mass in grams, c is specific heat in joules per gram degree, or $J/g \cdot °C$, and ΔT is the temperature change, or final temperature minus initial temperature. The Greek letter delta, Δ, indicates a change in a measured value, and always means final value minus initial value: $\Delta X = (X_f - X_i)$.

Specific heat is a property of a pure substance. It is the number of joules of energy that are required to raise the temperature of 1 g of the substance by 1°C. The specific heat of water is $4.18 \ J/g \cdot °C$. This value is used often in calorimetry experiments; it is one you should remember.

When a "hot" object comes into contact with a "cold" object — that is, when an object at higher temperature comes into contact with an object at lower temperature — heat flows from the hot object to the cold object. The hot object raises the temperature of the cold object, and the cold object cools the hot object. Eventually they reach the same intermediate temperature.

Heat flow between objects can be measured in a **calorimeter.** A perfect calorimeter is an isolated segment of the universe that allows no heat to flow to or from its contents during an experiment. It follows from the law of conservation of energy that, in a perfect calorimeter,

$$\Sigma Q = 0 \qquad (9.2)$$

where ΣQ is the sum of all the individual changes in heat content within the calorimeter. We will assume that the calorimeters used in this experiment are "perfect." This means that any heat transferred between

* The joule is still in the process of replacing the calorie as the standard heat unit in chemistry, and the calorie is still in common use. The calorie is now defined as exactly 4.184 joules.

its contents and the surroundings or between its contents and the calorimeter itself are negligible and may be disregarded.

In this experiment, you will measure the specific heat of two metals, one known and the other unknown. The known metal will be copper, which has a specific heat of 0.38 J/g · °C. You will calculate the percent error from this known value. The same experimental method will be used to find the specific heat of the unknown metal.

The laboratory procedure is to heat a weighed metal sample in boiling water until it reaches the temperature of the water. This is the initial temperature of the metal. The mass and initial temperature of water in a "coffee cup" calorimeter are measured. The metal is placed into the calorimeter, and the final temperature reached by both is recorded. According to Equation 9.2, the heat flow of the metal plus the heat flow of the water add up to zero:

$$Q_w + Q_m = 0 \tag{9.3}$$

$$Q_w = -Q_m \tag{9.4}$$

What Equation 9.4 says is that the heat lost by the metal is gained by the water. Substituting the expressions of Equation 9.1 into Equation 9.4 gives

$$m_w \times c_w \times \Delta T_w = -(m_m \times c_m \times \Delta T_m) \tag{9.5}$$

All values are known except the specific heat of the metal. Note the negative sign on the right side of the equation.

The specific heat of the unknown metal will be used to estimate the atomic mass of the element, and from that you can guess its identity. This will be done with the Law of Dulong and Petit, which was proposed in 1819. In essence, this law says that the product of the specific heat and atomic mass of all solid elements is equal to 26. From this,

$$\text{Atomic mass} = \frac{26}{\text{specific heat}} \tag{9.6}$$

SAFETY PRECAUTIONS

No hazardous chemicals are used in this experiment — only hazardous temperatures. You will be working with fairly large quantities of boiling water, pouring from one container to another, and dropping pieces of metal into boiling water. Do it carefully, anticipating and avoiding the conditions that might cause you to lose control of a beaker and burn yourself or others as a result. Also in this experiment you will be working with Bunsen burners. If you have long hair, be sure to tie it back.

PROCEDURE

Note: Record all mass measurements in grams to the nearest 0.01 g. Record all temperature measurements in degrees Celsius to the nearest 0.1°C. The smallest graduation on your thermometer must be no larger than 0.2°C for satisfactory results on this experiment.

1. PREPARATION FOR CALORIMETER RUNS

A) Place about 500 mL of tap water into a beaker. Set it aside so it will come to room temperature. *Do not place it near an operating Bunsen burner.* This will be your source of calorimeter water throughout the experiment.

B) Fill a beaker, 600 mL or larger, to about 80 percent of capacity with deionized water. (The deionized water is not required from a purity standpoint, but its use avoids the buildup of hard-to-remove scale that forms when tap water is boiled in a glass vessel.) Mount the beaker over wire gauze on a ring stand and heat to boiling. Proceed to Parts C and D while waiting.

From time to time in the experiment you will have to replace the water that has boiled out of the beaker with other water *already at the boiling temperature.* If some "community" source of boiling water is not available, establish your own by setting up a second beaker in the same manner as the first.

C) Select a piece of copper and an unknown metal sample from among those provided. Record the identification number of the unknown on your data sheet. Measure the mass of each piece and record it in grams to the nearest centigram.

D) Using crucible tongs, place each piece of metal in the water being heated. Position them in such a way that they will be easy to pick up and transfer to the calorimeter later. Be careful not to drop the metal in such a manner that it hits the bottom with sufficient force to break the beaker. The metal piece should remain in the boiling water for at least 30 minutes. As you proceed with the next steps, keep watch over the water level, which must be high enough to cover the metal completely. If too much water boils away, replenish it from the community source of boiling water, or from your second beaker, as explained in B above.

E) When the water in the heating bath is boiling, determine and record its temperature to the nearest 0.1°C. Set the thermometer aside and allow it to cool to room temperature. Two assumptions are made about the temperature of the boiling water: first, that it will remain constant throughout the experiment; and second, that the metal immersed in it will reach the same temperature, which is the initial temperature of the metal for later calculations.

F) Select and prepare a place on the laboratory desk for your calorimeter runs. It should be as far as possible from an operating Bunsen burner from which it might absorb heat, but not so far that the metal loses temperature while being transferred from the boiling water to the calorimeter. It is convenient to place a split rubber stopper on your thermometer (see Figure LP–6 in the Laboratory Procedures section) and clamp it on a ring stand, or to place your thermometer in a buret clamp, as depicted in Figure 9–1.

2. FIRST CALORIMETER RUN

A) Select a polystyrene coffee cup (preferably two, one nested in the other) for your calorimeter. Weigh the cup(s) to the nearest centigram. Pour about 100 mL of your calorimeter water from Step 1A into the cup and weigh again to determine accurately the mass of water in the cup. *Do not attempt to make it exactly 100 mL.* Record the data.

B) Place the thermometer into the calorimeter water, as shown in Figure 9–1. When the temperature has remained constant for about 1 minute, record that temperature to the nearest 0.1°C. In this and all other temperature measurements, be sure the thermometer bulb is totally immersed in the water.

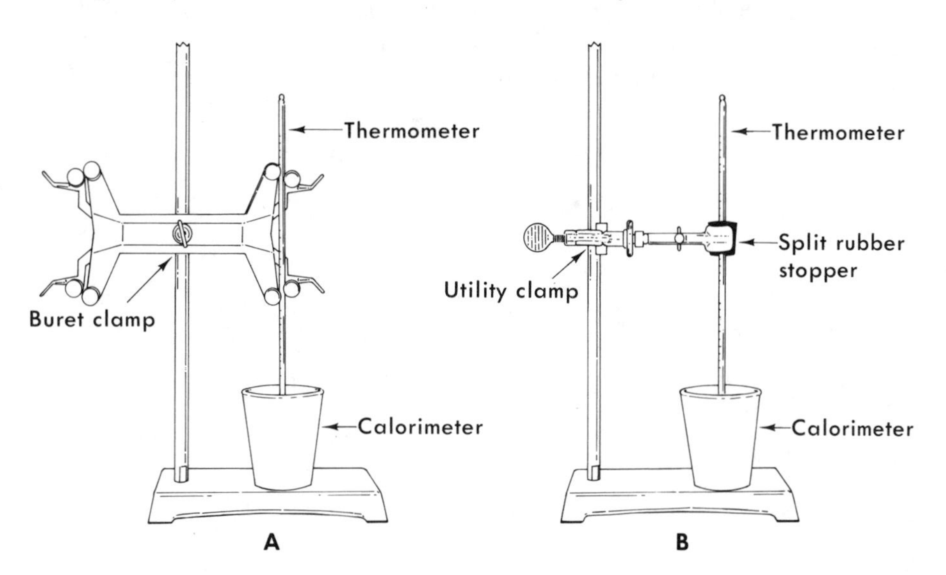

Figure 9–1

C) *(The next step is critical. The period in which the metal is out of water must be held to an absolute minimum to keep heat loss to the air as low as possible.)* Using crucible tongs, lift the piece of copper from the boiling water and hold it above the water level but below the top of the beaker for about 2 seconds to allow the boiling water to drain off. Then, *as quickly as possible,* place the metal into the calorimeter water. Be sure it is completely submerged. Be careful not to splash water out of the calorimeter. If this does occur, return the metal to the boiling water and start again at Step 2A.

D) Continuously and gently stir the water with a glass stirring rod, being careful not to break the bulb of the thermometer. Watch the thermometer. Read and record to the nearest 0.1°C the highest temperature that is reached before it begins to drop again. This may take several minutes.

E) Pour the water from the calorimeter into the sink. Place the copper back into the boiling water to be heated for a second run.

3. SUBSEQUENT CALORIMETER RUNS

Again pour about 100 mL of water into your calorimeter and weigh it to the nearest centigram. Repeat Steps 2B through 2E, using the unknown metal. The metal should be used alternately in subsequent runs, giving each sample enough time in the boiling water to reach its temperature. Make at least two runs with your copper sample and two with the unknown. At the end of the experiment, dry the metal pieces and return them to the place indicated by your instructor.

CALCULATIONS

Calculate the specific heat of the metal for each run of the experiment, using Equation 9.5, in which the specific heat is the only unknown. Find the percent error for each run with the copper, using the equations

$$\text{Percentage error} = \frac{\text{error}}{\text{accepted value}} \times 100$$

$$\text{Percentage error} = \frac{|\text{experimental value} - \text{accepted value}|}{\text{accepted value}} \times 100$$

Find the average specific heat for the two runs with the unknown metal. Estimate the atomic mass of the metal by substitution into Equation 9.6. If your estimated atomic mass is less than 100, assume your result to be correct to ± 4 in the doubtful digit and record the *range* of atomic masses indicated by your experiment. For example, if your experimental value of atomic mass is 38, the indicated range is from 34 to 42. Knowing your range of atomic masses, try to identify the element that is your unknown metal. Using the 34 – 42 atomic mass range, your element could be chlorine, argon, potassium, or calcium. Other things you know about those elements should help you in selecting the most likely one as your unknown.

If your estimated atomic mass is more than 100, assume the result to be correct to ± 1 in the doubtful digit and proceed as above.

Experiment 9
Work Page

NAME ....

DATE .... SECTION ....

DATA

Run number					
Metal or unknown No.					
Mass of metal (g)					
Temperature of boiling water (°C)					
Mass of calorimeter (g)					
Mass of calorimeter + water (g)					
Initial temperature of calorimeter water (°C)					
Final temperature (°C)					

CALCULATIONS AND RESULTS

Record all results in the table on the following page. Show below and on the next page all calculations for one column of data and results.

CALCULATIONS AND RESULTS

Mass of water (g)					
ΔT of water (°C)					
ΔT of metal (°C)					
Heat flow (J) (absolute value)					
Specific heat (J/g · °C)					
Percent error (copper only)					
Estimated atomic mass range for unknown (amu)					
Unknown element					

Experiment 9

Report Sheet

NAME ....

DATE .... SECTION ....

DATA

Run number					
Metal or unknown No.					
Mass of metal (g)					
Temperature of boiling water (°C)					
Mass of calorimeter (g)					
Mass of calorimeter + water (g)					
Initial temperature of calorimeter water (°C)					
Final temperature (°C)					

CALCULATIONS AND RESULTS

Record all results in the table on the following page. Show below and on the next page all calculations for one column of data and results.

CALCULATIONS AND RESULTS

Mass of water (g)					
ΔT of water (°C)					
ΔT of metal (°C)					
Heat flow (J) (absolute value)					
Specific heat (J/g · °C)					
Percent error (copper only)					
Estimated atomic mass range for unknown (amu)					
Unknown element					

Experiment 9

Advance Study Assignment

NAME ....

DATE .... SECTION ....

1) A student places 138 g of an unknown metal at 99.6°C into 60.50 g of water at 22.1°C. The entire system reaches a uniform temperature of 31.6°C. Calculate the specific heat of the metal.

2) If the actual specific heat of the metal in Problem 1 is 0.25 J/g · °C, calculate the percentage error.

3) If, as in the example in the Calculations section, your unknown element has an atomic mass in the range of 80 to 88, which element is it most apt to be? Justify your choice.

Experiment 10

Chemical Names and Formulas: A Study Assignment

Performance Goal

10–1 Within the limits discussed in this exercise, and using a periodic table for reference, given the name (or formula) of any chemical species among the classifications below, write the formula (or name):

Elements in their stable form
Molecular binary compounds
Binary acids; oxyacids
Monatomic ions; polyatomic ions
Ionic compounds

INTRODUCTION

This study assignment presents a brief summary of the rules for writing formulas and naming substances commonly encountered in an introductory chemistry course. Basic definitions are stated, but theory relating to chemical bonding and the formation of ions is not considered. The purpose of this exercise is to practice writing formulas and names with help immediately available to clear up points that you may not understand. Hopefully you will *master* formula writing techniques during this laboratory period.

CHEMICAL OVERVIEW

Elements

This discussion will be limited to the more common elements listed in Figure 10–1. Given the name of one of these elements, you should be able to write its symbol, using a full periodic table for reference; given the symbol, you should be able to identify the element by name. This requires a certain amount of memorization, but the sheer memory work is lessened if you relate elemental names and symbols to the periodic table.

Generally the chemical formula of an element in its stable form at room conditions is simply the symbol of the element. Seven gaseous elements, however, are not stable as individual atoms; two atoms combine to form a **diatomic molecule** as the unit particle of each of those elements. These elements and their correct chemical formulas are nitrogen, N_2; oxygen, O_2; hydrogen, H_2; fluorine, F_2; chlorine, Cl_2; bromine, Br_2; and iodine, I_2. Always remember to include the subscript 2 when writing the formulas of these substances *as elements, uncombined with any other elements.* In particular, notice that *this has nothing to do with these elements as they exist in compounds.* This is evident in the formulas for water, H_2O, and dinitrogen trioxide, N_2O_3.

Molecular Binary Compounds

Compounds made up of atoms held together entirely by covalent bonds are called **molecular compounds.** When a compound consists of two kinds of elements, it is called a **binary compound.** Molecular binary

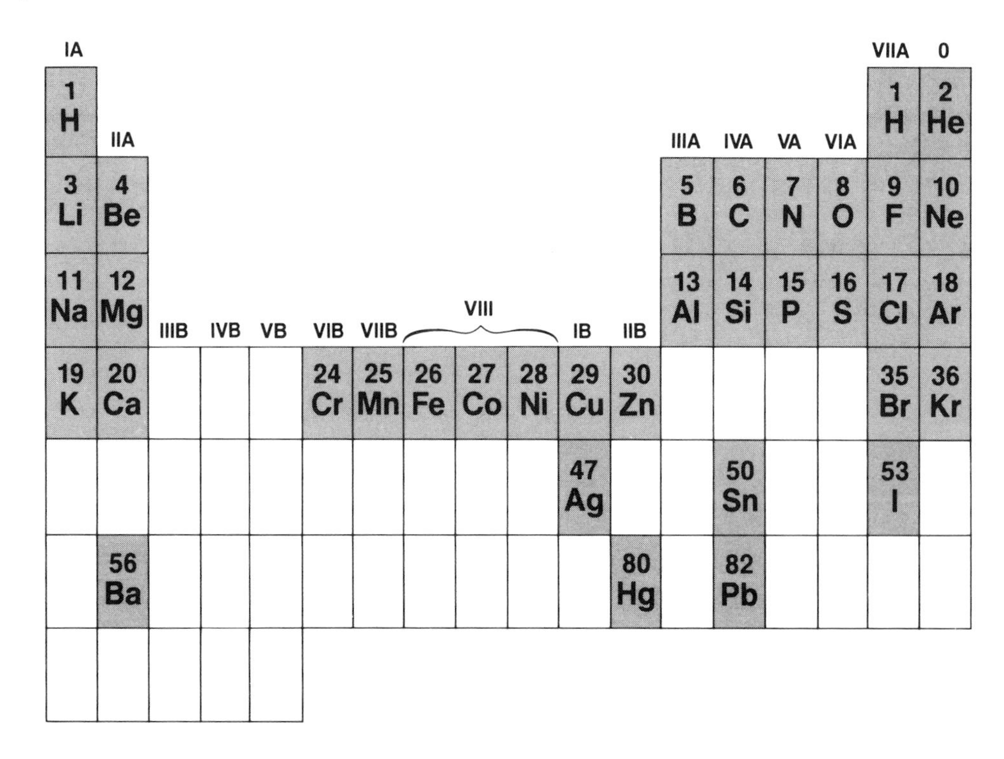

Figure 10–1. Partial periodic table showing the symbols and locations of the more common elements. The symbols above and the list that follows identify the elements you should be able to recognize or write, referring only to a complete periodic table. Associating the names and symbols with the table makes learning them much easier. The elemental names are:

aluminum	bromine	chromium	iodine	magnesium	nitrogen	silver
argon	calcium	copper	iron	manganese	oxygen	sodium
barium	carbon	fluorine	krypton	mercury	phosphorus	sulfur
beryllium	chlorine	helium	lead	neon	potassium	tin
boron	cobalt	hydrogen	lithium	nickel	silicon	zinc

compounds therefore consist of two elements held together by covalent bonds. These elements are generally *both nonmetals.* You may use this identification feature to distinguish molecular binary compounds from ionic binary compounds that will be discussed shortly.

Molecular binary compounds are identified by names consisting of two words. The main part of the first word is simply the name of the element appearing first in the formula; the main part of the second word is the name of the element appearing second in the formula, modified by an *-ide* suffix. The other part of each word is a prefix indicating the number of atoms of that particular element in the molecule. This is illustrated in the name dinitrogen trioxide for N_2O_3, in which *di-* is the prefix for 2 and *tri-* is the prefix for 3. A list of prefixes for numbers from 1 to 10 is given in Table 10–1. When the molecule contains only one atom of an element, the prefix *mono-* is frequently omitted, unless the species named is one of two or more compounds formed from the same two elements, such as CO, carbon monoxide, as compared to CO_2, carbon dioxide.

Table10–1 Prefixes Used in Naming Covalent Binary Compounds

mono-	= 1	hexa-	= 6
di-	= 2	hepta-	= 7
tri-	= 3	octa-	= 8
tetra-	= 4	nona-	= 9
penta-	= 5	deca-	= 10

Acids

Inorganic acids, and some organic acids, are compounds that yield a hydrogen ion, or proton, when they ionize. (A proton and a hydrogen ion are the same thing. A hydrogen atom consists simply of a proton and an electron. When the electron is removed, producing a hydrogen ion, the only thing left is the proton.) Formulas of such acids are written with the ionizable hydrogen appearing first. This feature can usually be used to identify a formula as that of an acid.

A **binary acid** consists of hydrogen and one other nonmetallic element, usually in water solution. A binary acid is named by surrounding the root of the nonmetal with the prefix *hydro-* and the suffix *-ic*. Thus HCl is hydrochloric acid, the *chlor* coming from chlorine. The name hydrosulfuric acid suggests that the element other than hydrogen is sulfur. Its formula is H_2S.

Oxyacids contain oxygen as well as hydrogen and another nonmetal. The name of the most common oxyacid of each nonmetal is the root of the nonmetal followed by *-ic*. Thus H_2SO_4 is sulfuric acid, and the formula for the common oxyacid of chlorine, called chloric acid, is $HClO_3$. These names and formulas are somewhat similar to the names and formulas of the hydro-ic acids. Catch the distinction: *hydro-ic* acids have no oxygen, whereas *-ic* acids do contain oxygen.

There are six so-called *-ic* acids whose names and formulas you should memorize, as they constitute the base from which we will develop our approach to learning the names and formulas of a large number of chemical compounds. If you memorize these six acids, plus some prefixes and suffixes, you will be able to figure out all the other names and formulas without further memorization. The six acids are:

chloric, $HClO_3$	carbonic, H_2CO_3
sulfuric, H_2SO_4	phosphoric, H_3PO_4
nitric, HNO_3	acetic, $HC_2H_3O_2$

Acetic acid is the best known of a large group of organic acids that contain hydrogen but ionize only slightly in water. Organic chemists write the formulas for such acids differently, but for the purpose of this exercise we will follow the usual procedure of writing the ionizable hydrogen first.

The number of oxygen atoms may vary in oxyacids of the same nonmetal. Chlorine, for example, forms four oxyacids: $HClO_4$, $HClO_3$, $HClO_2$, and HClO. The names of these compounds are distinguished from each other by a series of prefixes and suffixes that are explained in Table 10–2. The key to the entire nomenclature system is the number of oxygen atoms *compared to the number in the -ic acid.* Study Table 10–2 to help you memorize these prefixes and suffixes and understand their use.

Table10–2 Names of Oxyacids and Oxyanions of Chlorine (HCl included for comparison)

I Acid Name	II Acid Suffixes and Prefixes	III Acid Formula	IV Oxygens Compared to -ic Acid	V Ion Name	VI Ion Suffixes and Prefixes	VII Ion Formula
hydrochloric (binary acid)	*hydro-ic*	HCl	no oxygen	chloride	*-ide* named as monatomic anion	Cl^-
hypochlorous	*hypo-ous*	HClO	−2	hypochlorite	*hypo-ite*	ClO^-
chlorous	*-ous*	$HClO_2$	−1	chlorite	*-ite*	ClO_2^-
CHLORIC	*-IC*	$HClO_3$	SAME	CHLORATE	*-ATE*	ClO_3^-
perchloric	*per-ic*	$HClO_4$	+1	perchlorate	*per-ate*	ClO_4^-

Nonmetals of the same chemical family frequently form acids that are similar in name and formula. Among the halogens, for example, HCl is hydrochloric acid, HF is hydrofluoric acid, HBr is hydrobromic acid, and HI is hydroiodic acid. The similarities extend to oxyacids for bromine and iodine, but not for fluorine, which forms no oxyacids. We thus find that $HBrO_2$ is bromous acid, and HIO_4 is periodic acid.

Aside from the halogens, only sulfur and nitrogen form important oxyacids other than their well known *-ic* acids. In both cases it is the *-ous* acid that is formed, each with one less oxygen atom than is present in the *-ic* acid. Thus HNO_2 is the formula for nitrous acid, and sulfurous acid has the formula H_2SO_3. Selenium and tellurium, atomic numbers 34 and 52, in the same column of the periodic table as sulfur, form corresponding *-ous* acids.

Oxidation State: Oxidation Number

Chemists use a set of **oxidation numbers,** or consider the **oxidation state** of an element, in discussing oxidation–reduction reactions. These numbers are also part of the modern nomenclature system. The rules by which these numbers are assigned are as follows.

1. The oxidation number of any elemental substance is zero.
2. The oxidation number of a monatomic ion is the same as the charge on the ion.
3. The oxidation number of combined oxygen is -2, except in peroxides (-1) and superoxides ($-\frac{1}{2}$). We will not encounter peroxides or superoxides in this assignment.
4. The oxidation number of combined hydrogen is $+1$, except in hydrides (-1).
5. In any molecular or ionic species, the sum of the oxidation numbers of all atoms in the species is equal to the charge on the species.

The manner in which these rules are applied will be discussed as the need arises.

Monatomic Ions

A monatomic ion is a single atom that has acquired an electrical charge by gaining or losing one, two, or three electrons. Its formula is the symbol of the element followed by a superscript indicating the charge. For example, the formula of a calcium ion is Ca^{2+}, and for a chloride ion, Cl^-. It is important that the charge be indicated for an ion. Without that charge, the formula would be that of an electrically neutral atom from which the ion was formed, a very different species with very different chemical properties. Ions with a negative charge are called **anions;** ions with a positive charge are called **cations.**

The nonmetals in Groups VA, VIA, and VIIA form monatomic anions by gaining electrons. Ions from Group VA elements have a $3-$ charge, as in N^{3-}; from Group VIA, a $2-$ charge, as in O^{2-}; and from Group VIIA, a $1-$ charge, as in F^-. The name of a monatomic anion is simply the name of the element, modified by an *-ide* suffix, as in nitride, oxide, or fluoride.

Metals in Groups IA, IIA, and IIIA form cations with charges of $1+$, $2+$, and $3+$, respectively. Many metals in the B groups of the periodic table form two monatomic ions that differ in charge. The best example is iron, which yields the Fe^{2+} and Fe^{3+} ions. These ions are distinguished by adding the oxidation state, or charge, to the name of the element. Accordingly, Fe^{2+} is the iron(II) ion, and Fe^{3+} is the iron(III) ion. Notice how these names are written; the oxidation state is written in Roman numerals *and enclosed in parentheses* immediately after the name of the element, with no space between the name and the parentheses.

Caution: Students often neglect to enclose the oxidation state in parentheses; the name is not correctly written if the parentheses are missing.

The names of iron(II) and iron(III) ions are pronounced "iron two" and "iron three" respectively.

Notice that oxidation states in the names of monatomic ions are used only to distinguish between ions of the same element that have different charges. Oxidation numbers are not commonly used if a metal forms only one kind of ion. The two monatomic ions of copper are an exception to this. The copper(II) ion, Cu^{2+}, is so much more common than the copper(I) ion, Cu^+, that the name "copper ion" is understood to apply to Cu^{2+}. Copper(I) must be used to identify the Cu^+; and you are always correct if you use copper(II) for Cu^{2+}.

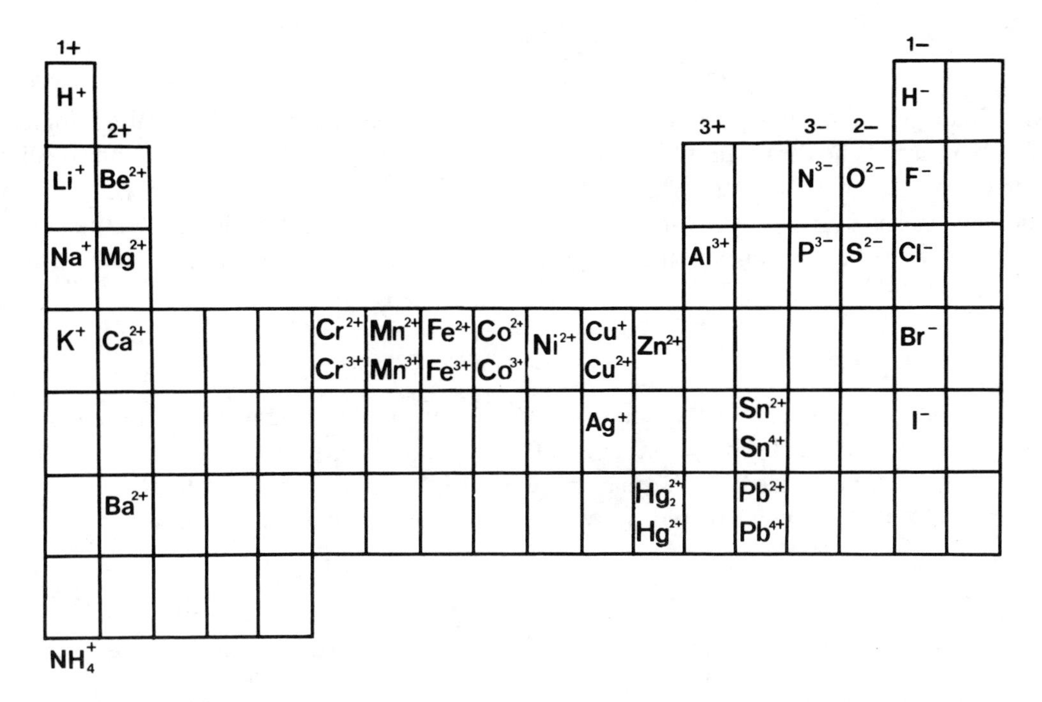

Figure 10–2. Partial periodic table of common ions. Notes: (1) Tin (Sn) and lead (Pb) form monatomic ions in a +2 oxidation state. In their +4 oxidation states they are more accurately described as being covalently bonded but such compounds are frequently named as if they were ionic compounds, (2) Hg_2^{2+} is a diatomic elemental ion. Its name is mercury(I), indicating a +1 charge from each atom in the diatomic ion, (3) ammonium ion, NH_4^+, is included as the only other common polyatomic cation, thereby completing this table as a minimum list of the cations you should be able to recall simply by referring to a full periodic table.

The cations formed by mercury require special comment. The mercury(II) ion, Hg^{2+}, is a typical monatomic ion. There is also a mercury(I) ion, but it is diatomic. Its formula is Hg_2^{2+}. The mercury(I) name is logical if you realize that *each atom* is contributing a 1+ charge to the diatomic ion.

Figure 10–2 locates in a periodic table the monatomic ions you should be able to recognize on sight, or write if given the name of the ion.

Polyatomic Anions Derived from the Total Ionization of Oxyacids

When an oxyacid ionizes, the resulting anion has more than one atom; it is a *polyatomic* anion. These are *oxyanions,* so called because they contain oxygen. Names of oxyanions are related to the acid from which they come; the prefix or suffix of the acid is replaced by a prefix or suffix for the anion. The system is illustrated for chlorine in Table 10–2, page 119. Memorize these prefixes and suffixes, and you will be able to apply them to a large number of compounds, including many you may never have heard of before.

The negative charge on an ion from the total ionization of an oxyacid is equal to the number of hydrogen atoms in the neutral acid molecule. Chloric acid, with one hydrogen, produces an oxyanion with a single negative charge, ClO_3^-; sulfuric acid, with two hydrogens, yields the double negative sulfate ion, SO_4^{2-}; and removal of three hydrogens from phosphoric acids yields an ion with a 3− charge, PO_4^{3-}.

Oxyanions Derived from the Stepwise Ionization of Polyprotic Acids

When an acid containing two or more hydrogen atoms ionizes, it loses the hydrogen ions one by one. There are, therefore, intermediate ions that contain hydrogen. The stepwise ionization of sulfuric acid may be represented by

$$H_2SO_4 \xrightarrow{-H^+} HSO_4^- \xrightarrow{-H^+} SO_4^{2-}$$

The HSO_4^- ion can be thought of as a sulfate ion with a hydrogen attached. It is given the logical name, hydrogen sulfate ion. When triprotic phosphoric acid, H_3PO_4, ionizes, there are two intermediate ions, $H_2PO_4^-$ and HPO_4^{2-}. The first of these is the phosphate ion with two hydrogens attached, so it is called the dihydrogen phosphate ion, which distinguishes it from HPO_4^{2-}, the hydrogen (or monohydrogen) phosphate ion. Intermediate ions from other polyprotic acids are named in a similar manner.

Other Polyatomic Ions

There are two other polyatomic ions that are so common you should recognize them instantly. These are the ammonium ion, NH_4^+, and the hydroxide ion, OH^-. Many other polyatomic ions exist, but it is not necessary that they be memorized at this time unless your instructor directs you to do so. Some of them are in Tables 10–3 and 10–4, which include most of the ions you are apt to encounter in a beginning chemistry course.

Table10–3 Common Cations

Ionic Charge: +1		*Ionic Charge: +2*		*Ionic Charge: +3*	
Alkali Metals: Group IA		*Alkaline Earths: Group IIA*		*Group IIIA*	
Li^+	Lithium	Be^{2+}	Beryllium	Al^{3+}	Aluminum
Na^+	Sodium	Mg^{2+}	Magnesium	Ga^{3+}	Gallium
K^+	Potassium	Ca^{2+}	Calcium	*Transition Elements*	
Rb^+	Rubidium	Sr^{2+}	Strontium	Cr^{3+}	Chromium(III)
Cs^+	Cesium	Ba^{2+}	Barium	Mn^{3+}	Manganese(III)
Transition Elements		*Transition Elements*		Fe^{3+}	Iron(III)
Cu^+	Copper(I)	Cr^{2+}	Chromium(II)	Co^{3+}	Cobalt(III)
Ag^+	Silver	Mn^{2+}	Manganese(II)		
Polyatomic Ions		Fe^{2+}	Iron(II)		
NH_4^+	Ammonium	Co^{2+}	Cobalt(II)		
		Ni^{2+}	Nickel		
		Cu^{2+}	Copper(II)		
Others		Zn^{2+}	Zinc		
H^+	Hydrogen	Cd^{2+}	Cadmium		
or		Hg_2^{2+}	Mercury(I)		
H_3O^+	Hydronium	Hg^{2+}	Mercury(II)		
		Others			
		Sn^{2+}	Tin(II)		
		Pb^{2+}	Lead(II)		

Ionic Compounds

Two rules govern the nomenclature of ionic compounds:

1. The name of an ionic compound is the name of the positive ion followed by the name of the negative ion.
2. The formula of an ionic compound is the formula of the positive ion followed by the formula of the negative ion, each taken as many times as may be necessary to bring the total charge to zero.

Table 10–4 Common Anions

Ionic Charge: −1				*Ionic Charge: −2*		*Ionic Charge: −3*	
Halogens: Group VIIA		*Oxyanions*		*Group VIA*		*Group VA*	
F^-	Fluoride	ClO_4^-	Perchlorate	O^{2-}	Oxide	N^{3-}	Nitride
Cl^-	Chloride	ClO_3^-	Chlorate	S^{2-}	Sulfide	P^{3-}	Phosphide
Br^-	Bromide	ClO_2^-	Chlorite	*Oxyanions*		*Oxyanion*	
I^-	Iodide	ClO^-	Hypochlorite	CO_3^{2-}	Carbonate	PO_4^{3-}	Phosphate
Acidic Anions		BrO_3^-	Bromate	SO_4^{2-}	Sulfate		
HCO_3^-	Hydrogen carbonate	BrO_2^-	Bromite	SO_3^{2-}	Sulfite		
HS^-	Hydrogen sulfide	BrO^-	Hypobromite	$C_2O_4^{2-}$	Oxalate		
HSO_4^-	Hydrogen sulfate	IO_4^-	Periodate	CrO_4^{2-}	Chromate		
HSO_3^-	Hydrogen sulfite	IO_3^-	Iodate	$Cr_2O_7^{2-}$	Dichromate		
$H_2PO_4^-$	Dihydrogen phosphate	NO_3^-	Nitrate	*Acidic Anion*			
Other Anions		NO_2^-	Nitrite	HPO_4^{2-}	Monohydrogen phosphate		
SCN^-	Thiocyanate	OH^-	Hydroxide	*Diatomic*			
CN^-	Cyanide	$C_2H_3O_2^-$	Acetate	O_2^{2-}	Peroxide		
H^-	Hydride	MnO_4^-	Permanganate				

To name an ionic compound when given the formula, you need only to recognize the ions present. You must be familiar with the number of oxygen atoms in the various oxyanions, as well as the rules by which the anions are named. For a compound having a cation from a metal that forms two different monatomic ions, you must apply the oxidation-state rules to determine which of those ions is present. If the compound is $FeCl_2$, for example, you must recognize that the chloride ion has a 1− charge. There are two chloride ions present, so that the total negative charge in the formula unit is 2−. The sum of all the oxidation numbers in the formula must be zero, which means the 2+ charge must come from the iron ion, and the compound must therefore be iron(II) chloride. Similar reasoning would lead to the conclusion that $FeCl_3$ is iron(III) chloride.

In writing the formulas of compounds in which a polyatomic ion appears more than once, the entire ion is enclosed in parentheses, followed by a subscript indicating the number of ions in the formula unit. For example, the formula of calcium nitrate is $Ca(NO_3)_2$. This is the only time parentheses are used. Specifically, they are not used when a polyatomic ion appears only once in the formula, as in calcium sulfate, $CaSO_4$. Nor is the symbol of a monatomic ion enclosed in parentheses just because it happens to have two letters, as in calcium bromide, $CaBr_2$.

Experiment 10

Report Sheet

NAME ....

DATE .... SECTION ....

General Instructions: For each substance whose name is given, write the formula; if the formula is given, write the name. Unless stated otherwise, a periodic table should be your only reference.

A. ELEMENTS

Write the formulas of the elements in their natural, stable states.

Iron	Na
Calcium	Cl_2
Nitrogen	Cu
Bromine	Mg
Potassium	Ni

B. MOLECULAR BINARY COMPOUNDS

Carbon dioxide	CBr_4
Dinitrogen tetroxide	CO
Iodine chloride	P_2O_3
Sulfur trioxide	SiS_2
Diphosphorus pentoxide	S_2F_6

C. ACIDS

Hydrobromic acid	HClO
Sulfuric acid	HI
Bromic acid	HNO_3
Phosphoric acid	H_2SO_3
Nitrous acid	HIO_4
Perchloric acid	$HBrO_2$

D. MONATOMIC AND POLYATOMIC IONS

Calcium ion	Fe^{2+}
Sulfate ion	Br^-
Hydrogen phosphate ion	ClO^-
Nitrite ion	CO_3^{2-}
Iron(III) ion	Cr^{3+}
Iodite ion	SO_3^{2-}
Sulfide ion	HCO_3^-

Experiment 10
Report Sheet

NAME ....

DATE .... SECTION ....

E. IONIC COMPOUNDS

Sodium nitrate	K_2SO_4
Calcium fluoride	Na_3PO_4
Potassium hydrogen sulfate	$Pb(NO_3)_2$
Sodium carbonate	$FeCl_3$
Potassium bromide	KIO_3
Iron(III) sulfide	$Ca(OH)_2$
Magnesium chloride	$Al_2(SO_4)_3$
Sodium dihydrogen phosphate	$HgCO_3$
Ammonium sulfate	$NaClO_2$
Copper(II) carbonate	KHS
Barium hydroxide	K_2O
Silver bromide	$NaHSO_3$
Mercury(II) sulfate	$(NH_4)_2CO_3$

E. IONIC COMPOUNDS (Continued)

Potassium nitrite	FeO
Calcium chlorate	$NaHCO_3$
Iron(II) hydroxide	CaI_2
Copper(I) iodate	NH_4Br
Aluminum sulfite	$BaCl_2$
Magnesium oxide	$FePO_4$
Lead(II) iodide	Ag_2SO_4
Sodium hypochlorite	$Co(OH)_2$
Lithium hydrogen sulfite	NH_4NO_2
Ammonium carbonate	Cu_2O
Mercury(I) chloride	K_3PO_4
Aluminum oxide	$(NH_4)_2HPO_4$
Potassium periodate	$AgBrO_3$

Experiment 10
Report Sheet

NAME ....

DATE .... SECTION ....

F. COMPOUNDS CONTAINING LESS COMMON IONS

(Refer to tables of cations and anions and the periodic table when writing these formulas.)

Strontium sulfate (strontium, atomic number 38)	
Cesium iodide (cesium, atomic number 55)	
Indium chloride (indium, atomic number 49)	
Tellurium trioxide (tellurium, atomic number 52)	
Calcium hydride	
Sodium cyanide	
Iron(III) thiocyanate	
Nickel chromate	

Experiment 10
Advance Study Assignment

NAME ....

DATE .... SECTION ....

1) Give the names and formulas of two oxyacids of sulfur.

2) Give an example of an ionic compound derived from: a) nitric acid, b) carbonic acid.

3) Write the formula of an ionic compound derived from:
   a) X^+ cation and Y^{3-} anion.

   b) X^{2+} cation and HY^- anion.

   c) X^{3+} cation and HY^{2-} anion.

Experiment 11

Chemical Equations: A Study Assignment

Performance Goal

11–1 Given information from which you can write the formulas for all reactants and all products for each of the following types of reactions, write the balanced chemical equation for the reactions:

Combination
Decomposition
Complete oxidation or burning of organic compounds
Ion combination, forming a precipitate or molecular product
Oxidation–reduction ("single replacement" equations only)
Other reactions in which reactants and products are identified

CHEMICAL OVERVIEW

A chemist uses a chemical equation to describe a chemical change. The general form of a chemical equation is

$$\text{Reactant 1} + \text{Reactant 2} + \ldots \rightarrow \text{Product 1} + \text{Product 2} + \ldots$$

The substances that enter into the reaction are called **reactants.** They are identified by their chemical formulas, written on the left side of the equation, and separated from each other by plus signs. The formulas of the new substances produced in the reaction, called **products,** are written on the right side, again separated by plus signs. The two sides of the equation are separated by an arrow pointing from the reactants to the products, indicating that the reactants are changed into the products. In reading a chemical equation, or expressing it in words, the arrow is frequently read as "yields," "produces," or "forms"; any other term that suggests the creation of a substance not originally present is equally satisfactory.

Symbols are frequently added to chemical equations to indicate the conditions under which the reaction occurs. The symbols (s), (ℓ), or (g) immediately after the formula of a substance indicate that the substance is in the solid, liquid, or gaseous state, respectively. A substance that is in aqueous (water) solution may have (aq) after its formula. An arrow pointing up after a formula indicates a gaseous product that escapes to the atmosphere; an arrow pointing down identifies a product that is precipitated, or formed into a solid, from a solution. Sometimes the arrow between reactants and products is lengthened, and words, formulas, temperatures, or other symbols are written above (or above and below) the arrow to indicate reaction conditions or other substances in the reaction vessel. None of these supplementary items will be used in this exercise; however, if your instructor requests that you use them, you should, of course, follow his or her directions.

A chemical equation does two things. First, it tells you what substances are involved in a chemical change. To do this accurately, it is essential that the substances be represented by their correct chemical formulas. It is assumed in this exercise that, given the name of a chemical, you are able to write its formula. Second, an equation has quantitative significance. It obeys the law of conservation of mass, which indicates that the total mass of all the reactants is equal to the total mass of all the products in an ordinary chemical change. In order for this to be true, the equation must have equal numbers of atoms of each individual element on the two sides of the equation. The equation is then said to be **balanced.**

These two characteristics of an equation lead to a simple two-step procedure by which an equation may be written:

1. Write the correct chemical formula for each reactant on the left and each product on the right.
2. *Using coefficients only,* balance the number of atoms of each element on each side of the equation. If no coefficient is written, its value is assumed to be one (1).

It is impossible to overemphasize the importance of following these two steps literally, and keeping them independent. In Step 1, write the correct formulas without concern about where the atoms come from, or how many atoms of an element may be present in some species on the other side of the equation. In Step 2, be sure that you balance the atoms of each element by placing whole-number coefficients in front of chemical formulas, and by no other means. Specifically,

DO NOT change a correct chemical formula in order to balance an element;
DO NOT add some real or imaginary chemical species to either side to make an element balance.

Quite often the word description of a chemical reaction will not identify all of the species that must be included in the equation. If you are familiar with the kinds of reactions described in the performance goal, you will be able to identify the substances not mentioned. The reaction types will be discussed as they are encountered.

EXAMPLES

The following examples are in the form of a program in which you learn by answering a series of questions. Obtain an opaque shield (a piece of cardboard, or a folded piece of paper you cannot see through) that is wide enough to cover this page. In each example place the shield on the book page so it covers everything beneath the first dotted line that runs across the page. Read to that point, and write in the space provided whatever is asked. Then lower the shield to the next dotted line. The material exposed will begin with the correct response to the question you have just answered. Compare this answer to yours, looking back to correct any misunderstanding if the two are different. When you fully understand the first step, read to the next dotted line and proceed as before.

A. Combination Reactions. A combination reaction occurs when two or more substances combine to form a single product. The reactants may be elements or compounds, perhaps one or more of each. Quite often the description of the reaction will give the chemical name of the product only. For example, the equation for the reaction in which sodium chloride is formed from its elements is $2\,Na + Cl_2 \rightarrow 2\,NaCl$. An example of a combination reaction between compounds is $CaO + H_2O \rightarrow Ca(OH)_2$.

Example 1

Write the equation showing how magnesium oxide is formed from its elements.

". . . magnesium oxide is formed . . ." indicates that magnesium oxide is the product of the reaction, so its formula will appear on the right side of the equation. ". . . from its elements" identifies magnesium and oxygen as the reactants whose formulas will be written to the left of the arrow. Complete Step 1 of the procedure by writing the unbalanced equation.

. . . .

1a. $Mg + O_2 \rightarrow MgO$

Remember, oxygen is a diatomic element; its correct formula is therefore O_2 and not simply O.

Step 2 calls for you to balance the atoms of each element on the two sides of the equation. As it stands, there is one magnesium atom on each side; magnesium is in balance. The left side of the equation has two atoms of oxygen, and the right side only one. What must you do to balance oxygen? Remember, there is only one way to do it, and watch out for the DON'Ts listed earlier. Balance the oxygen.

$$Mg + \quad O_2 \rightarrow \quad MgO$$

. . . .

1b. $Mg + O_2 \rightarrow 2\ MgO$

Remember, your only way to balance atoms of an element in Step 2 is to use coefficients in front of substances in the unbalanced equation. Some common WRONG responses to the above — and what is wrong with them — are:

$Mg + O_2 \rightarrow MgO + O$	There are no oxygen atoms in the reaction. This violates the second DON'T.
$Mg + O_2 \rightarrow MgO_2$	MgO_2 happens to be a real substance, but it is *not* the product of this reaction. This violates the first DON'T.
$Mg + O_2 \rightarrow Mg\ 2\ O$	Mg 2 O is not a chemical formula. Coefficients are placed in front of a formula, not in the middle, and they affect the entire formula.

The last of the three wrong balancing methods points out that in balancing oxygen we have *un*balanced magnesium. There is now one magnesium atom on the left and two on the right. Correct this now.

. . . .

1c. $2\ Mg + O_2 \rightarrow 2\ MgO$

There is another way you might have balanced $Mg + O_2 \rightarrow MgO$. You could have introduced the fractional coefficient 1/2 in front of oxygen: $Mg + 1/2\ O_2 \rightarrow MgO$. While fractional coefficients are technically correct, they are not commonly used, and should not be used in this exercise. They may be used as a means to the final equation, however. If you do choose to balance the equation with a fractional coefficient, as above, you can then multiply the entire equation by 2 (doubling *each* coefficient), giving $2\ Mg + O_2 \rightarrow 2\ MgO$. Incidentally, equations should be written with the *smallest* whole-number coefficients. If, in your balancing procedure, you happened to arrive at $4\ Mg + 2\ O_2 \rightarrow 4\ MgO$, you could divide the entire equation — each coefficient — by 2 to get the desired result.

Example 2

Write the equation for the formation of iron(III) oxide from its elements.

Complete the first step by writing the formulas of the reactants on the left and the formula of the product on the right.

. . . .

2a. $Fe + \quad O_2 \rightarrow \quad Fe_2O_3$

Start with the iron; balance it first and leave oxygen unbalanced.

. . . .

2b. $2\ Fe + \quad O_2 \rightarrow \quad Fe_2O_3$

There are two thought processes by which balancing may be completed, both leading to the same result. Both will be discussed — after you have balanced the rest of the equation yourself.

. . . .

2c. $4\ Fe + 3\ O_2 \rightarrow 2\ Fe_2O_3$

Oxygen atoms come two to the package in O_2 molecules, and three to the package in Fe_2O_3 units. Six atoms — 2 times 3 — is the smallest number of atoms by which a 3-and-2 combination can be equalized. If you take 3 packages of 2 each, you will have the same number as 2 packages of 3 each. This fixes the coefficients of O_2 and Fe_2O_3. The coefficient of iron is adjusted to correspond with the iron atoms in 2 Fe_2O_3.

A second way of reaching the final equation is to select the fractional coefficient of O_2 that will give the proper number of oxygen atoms to balance the three on the right side of $2\ Fe + O_2 \rightarrow Fe_2O_3$. With three oxygens on the right, we need three on the left, where they come two to a package in O_2. We therefore need $1\frac{1}{2}$ packages, or 3/2, yielding $2\ Fe + 3/2\ O_2 \rightarrow Fe_2O_3$. This balanced equation can be cleared of fractions by multiplying all coefficients by 2, giving $4\ Fe + 3\ O_2 \rightarrow 2\ Fe_2O_3$.

It is worthwhile to become familiar with both methods. The 3-and-2 combination appears frequently enough to justify the routine 2-of-3 and 3-of-2 thought process. It is convenient to realize that if you need *X* atoms of oxygen from O_2 molecules, the number of molecules required is *X*/2. Doubling the equation yields whole-number coefficients.

B. Decomposition Reactions. The chemical change in which a single reactant decomposes into two or more products is a decomposition reaction. This is just the opposite of a combination reaction; indeed, many combination reactions can be reversed, as $2\ NaCl \rightarrow 2\ Na + Cl_2$. The reaction $2\ Al(OH)_3 \rightarrow Al_2O_3 + 3\ H_2O$ illustrates a decomposition of a compound into two simpler compounds. Another type of decomposition reaction occurs when hydrates (compounds containing water of hydration) are heated. $Na_2CO_3 \cdot 10H_2O \xrightarrow{\Delta} Na_2CO_3 + 10\ H_2O$ illustrates such a reaction, where Δ written over the arrow generally means "applying heat."

Example 3

Calcium carbonate is decomposed into calcium oxide and carbon dioxide by heat. Write the equation.

The first step is to write the formulas of reactants and products in their proper places. Proceed that far.

. . . .

3a. $CaCO_3 \rightarrow \quad CaO + \quad CO_2$

Now Step 2: balance the atoms of each element on the two sides of the equation.

. . . .

3b. $CaCO_3 \rightarrow CaO + CO_2$

Sometimes balancing an equation is easy — particularly when all coefficients are 1!

C. Complete Oxidation or Burning of Organic Compounds. Other than the oxides of carbon, carbonates, and a few other substances, the compounds of carbon are classified as **organic compounds.** Hydrogen is almost always present in an organic compound, and oxygen is a third very common element. When compounds containing carbon and hydrogen, or carbon, hydrogen, and oxygen, react *completely* with an excess of oxygen, the products are always carbon dioxide and water. Such a reaction may occur with the oxygen in the air, giving heat and light, in which case the process is called **burning;** and it may occur in living organisms, again giving off heat and other forms of energy, in which case it is referred to as **oxidation.** The description of such a reaction may be very brief: Compound *X* is completely oxidized, or Compound *Y* is burned in air. In both reactions you must recognize oxygen as an unnamed reactant to be included in the equation, and write the formulas of carbon dioxide and water as the products. $2\ C_6H_{14} + 19\ O_2 \rightarrow 12\ CO_2 + 14\ H_2O$ is an example of a burning reaction. Because organic compounds are frequently quite large, equations may have large coefficients; but don't let that bother you, as they are reached by the same method outlined above.

Example 4

Write the equation for the complete oxidation of methyl ethyl ketone, $CH_3COC_2H_5$.

Methyl ethyl ketone has been chosen for this example because its equation includes all the little things you must look out for in writing oxidation equations. First, notice that organic chemists sometimes write formulas in ways that seem strange to the beginning student. This is because the sequences of elements and certain combinations in the formula suggest how atoms are arranged in the molecule and identify the kind of compound it is. Secondly, you must be sure to count *all* the atoms of a given element in a molecular formula when balancing, such as 4 carbon atoms, 8 hydrogen atoms, and 1 oxygen atom in a molecule of $CH_3COC_2H_5$. A third point will show up later. Right now, complete Step 1 by writing the formulas of reactants and products in their proper places in an unbalanced equation.

. . . .

4a. $CH_3COC_2H_5 + \quad O_2 \rightarrow \quad CO_2 + \quad H_2O$

Always remember that, although it is unnamed in the statement of the reaction, oxygen is a second reactant and the products are carbon dioxide and water.

To begin Step 2, you balance both carbon and hydrogen. With the warning already given, add those coefficients to the equation.

. . . .

4b. $CH_3COC_2H_5 + \quad O_2 \rightarrow \quad 4\ CO_2 + \quad 4\ H_2O$

With carbon and hydrogen balanced, and oxygen in its elemental form on the left, oxygen can be balanced simply by placing in front of oxygen the coefficient that does the job. Sounds simple, but be careful . . .

4c. $2\ CH_3COC_2H_5 + 11\ O_2 \rightarrow 8\ CO_2 + 8\ H_2O$

Starting from $CH_3COC_2H_5 + O_2 \rightarrow 4\ CO_2 + 4\ H_2O$, you count 12 oxygen atoms on the right side of the equation. On the left, *one of the required 12 oxygen atoms comes from the reactant,* and the remaining *eleven* come from O_2. This is the third thing you must look out for in balancing oxidation equations, being sure not to overlook oxygen present in the compound being oxidized. With 11 oxygen atoms to come from O_2, you can balance the equation with a fractional coefficient: $CH_3COC_2H_5 + 11/2\ O_2 \rightarrow 4\ CO_2 + 4\ H_2O$. Doubling the entire equation gives whole-number coefficients, as required.

D. Oxidation–Reduction Reactions. The words **oxidize** and **oxidation** have meaning in chemistry other than "reaction with oxygen," as suggested in the foregoing section. In its broader meaning, oxidation means loss of electrons. If one reactant loses electrons, another reactant must gain those electrons. The process of gaining electrons is called **reduction.** A reaction in which oxidation and reduction occur — and they must always occur simultaneously — is called an **oxidation–reduction reaction,** frequently shortened to "redox" reaction.

In this exercise we will be concerned with only one kind of redox reaction. The equation has the appearance of an element reacting with a compound in such a manner that the element replaces one of the elements in the compound. $Zn + Cu(NO_3)_2 \rightarrow Cu + Zn(NO_3)_2$ is such a reaction. It appears as if elemental zinc has replaced copper from $Cu(NO_3)_2$. This kind of equation is frequently called a **single replacement equation.** Given an element and an ionic compound as reactants, you should recognize the possibility of a redox reaction and be able to write the single replacement equation for that reaction. Whether or not the reaction actually occurs requires laboratory confirmation, of course.

Example 5

Gaseous hydrogen is released when zinc reacts with hydrochloric acid. Write the equation for the reaction.

The reactants and one of the products are identified. As you write the unbalanced equation (Step 1) for these three species, see if you can recognize the single replacement character of that equation and then figure out the formula of the second product.

5a. $Zn + \quad HCl \rightarrow \quad H_2 + \quad ZnCl_2$

In the reaction zinc is releasing, or replacing, hydrogen in HCl. The second product is therefore zinc chloride.

Balancing the equation is straightforward . . .

5b. $Zn + 2\ HCl \rightarrow H_2 + ZnCl_2$

Example 6

Write the equation for the reaction between aluminum and nickel nitrate.

This time you are given only the names of two reactants. Write their formulas on the left side of the arrow, leaving the product side blank.

6a. $Al + Ni(NO_3)_2 \rightarrow$

Here's where your skill in recognizing the possibility of a redox reaction comes into play. What possible products could come from these reactants? There is no indication that the nitrate ion decomposes. Ions of aluminum and nickel are both positively charged, so there is no way they could form a compound. The nitrate ion is negatively charged, so it could form a compound with an aluminum ion. It all points to a single replacement equation in which aluminum bumps nickel out of the compound. Complete Step 1 by writing the formulas of the products on the right side of the equation.

6b. $Al + \quad Ni(NO_3)_2 \rightarrow \quad Ni + \quad Al(NO_3)_3$

This example gives us an opportunity to introduce an important technique in balancing equations. A quick glance shows that aluminum and nickel are balanced, but nitrogen and oxygen are not. You could balance them individually, but there is an easier way. It was noted above that the *nitrate ion* does not decompose in the reaction; in other words, the nitrate ion is the same on the product side of the equation as it is on the reactant side. Any time a polyatomic (many atom) ion is unchanged in a chemical reaction, that ion may be balanced *as a unit* in the equation. In other words, your thought process should be, "There are two nitrate ions on the left, and three nitrate ions on the right. How do I balance them?" How *do* you balance a 3-and-2 combination? You already know that, so go ahead. While you're at it, be sure to do whatever is necessary to keep the aluminum and nickel in balance.

6c. $2\ Al + 3\ Ni(NO_3)_2 \rightarrow 3\ Ni + 2\ Al(NO_3)_3$

You need 3 nickel nitrate units, where nitrate ions appear in packages of two, to balance 2 aluminum nitrate units, where the nitrate ions appear in packages of three, giving 6 nitrate ions on each side of the equation. The coefficients for the metals complete the equation.

E. Ion Combination Reactions. As the name suggests, ion combination reactions involve the combination of ions from different sources to form a new product. If two ions are to combine, one must have a positive charge and the other must have a negative charge. The combination of a lead ion from lead(II) nitrate and a chloride ion from sodium chloride to form lead(II) chloride is a good example: $Pb(NO_3)_2 + 2\ NaCl \rightarrow PbCl_2 + 2\ NaNO_3$. If you look at the equation, it appears as if the positive and negative ions in the two reactants have simply "changed partners" in the products; the positive ion of the first reactant has joined up with the negative ion of the second reactant, and the negative ion of the first has combined with the positive ion of the second. This kind of equation is called a **double displacement equation.** Whenever you see an equation with two ionized reactants, you can make an intelligent prediction that the products will be derived from an exchange of ions, and write their formulas accordingly.

Most ion combination reactions occur in water solution. One of the driving forces for these reactions is the formation of an insoluble ionic solid, called a **precipitate.** In the example above, lead(II) chloride is insoluble in water, so it precipitates as the ions combine with each other.

The other driving force that brings ions together is the formation of a molecular product, in which a covalent (shared electron pair) bond forms between the reacting ions. The most common molecular product is water, as in $HCl + KOH \rightarrow KCl + H_2O$. This kind of reaction, in which an acid reacts with a base, is called a **neutralization reaction.** The ionic product formed (KCl, in this example) is classified as a **salt.**

Example 7

Write the equation for the precipitation reaction that occurs when sodium hydroxide is added to a solution of copper(II) nitrate.

You are given the reactants. Write their formulas to the left of the arrow, leaving the right side blank.

. . . .

7a. $NaOH + Cu(NO_3)_2 \rightarrow$

Once the reactant formulas are written, it is easier to see the new combinations of ions that are possible. Write the formulas of the product species on the right.

. . . .

7b. $NaOH + \quad Cu(NO_3)_2 \rightarrow \quad NaNO_3 + \quad Cu(OH)_2$

The positively charged sodium ion from the first reactant is shown as combined with the negatively charged nitrate ion from the second reactant, and the cation from the second combines with the anion from the first. The ions have "changed partners."

In balancing the equation, remember what you have learned about balancing as units any polyatomic ions that are unchanged in the reaction. In this case it is a lot easier than balancing oxygen by itself, appearing, as it does, in all four compounds. Go ahead — complete the equation.

. . . .

7c. $2\ NaOH + Cu(NO_3)_2 \rightarrow 2\ NaNO_3 + Cu(OH)_2$

In addition to the sodium and copper being balanced, the equation shows two hydroxide ions and two nitrate ions on each side.

Example 8

Write the equation for the neutralization of sulfuric acid by lithium hydroxide.

The procedure in this example is the same as in the last. This time complete Step 1, writing formulas of both reactants and products in their proper places.

. . . .

8a. $H_2SO_4 + \quad LiOH \rightarrow \quad Li_2SO_4 + \quad H_2O$

Balancing this equation is easy enough, but the thought processes are important. See if you can think it through in terms of hydrogen and hydroxide ions.

. . . .

8b. $H_2SO_4 + 2\ LiOH \rightarrow Li_2SO_4 + 2\ H_2O$

Lithium and the sulfate ions balance in the usual manner. The two hydrogens in H_2SO_4 are balanced by the first hydrogen (underlined) in 2 $\underline{H}OH$ (H_2O) molecules. The two hydroxide ions in 2 LiOH are balanced by the OH parts of 2 $H\underline{OH}$ molecules.

Water is not the only molecular product that may be formed, as the following example illustrates.

Example 9

Write the equation for the reaction between nitric acid and sodium acetate, $NaC_2H_3O_2$.

First the formulas of the reactants only . . .

. . . .

9a. $HNO_3 + \quad NaC_2H_3O_2 \rightarrow$

The formula of sodium acetate was given to you partly because the acetate ion is probably not familiar to you, but more importantly because we didn't want you to look it up. Instead, you should figure out what the charge on the ion is from the formula of sodium acetate. Then figure out how many hydrogen ions will combine with how many acetate ions to form the molecular product, acetic acid. Write that formula, as well as the formula of the salt produced in the reaction, on the right side of the equation. Balance the equation, too.

. . . .

9b. $HNO_3 + NaC_2H_3O_2 \rightarrow HC_2H_3O_2 + NaNO_3$

If there is one Na^+ ion in a formula unit of $NaC_2H_3O_2$, the acetate ion must consist of everything else in the compound, with a negative charge equal to the positive charge of the sodium ion: $C_2H_3O_2^-$. The 1− charge of the acetate ion requires the 1+ charge of a hydrogen ion, H^+, to form a neutral molecule, so the formula of acetic acid must be $HC_2H_3O_2$. The equation is balanced with all coefficients equal to 1.

F. Other Reactions in Which Reactants and Products Are Identified. Many kinds of chemical reactions do not fit into any of the classifications we have considered. Therefore, you cannot be expected to predict what substances might be formed in such reactions, nor to predict the formulas of these products. If you are given the names and/or formulas of all species in an unclassified reaction, however, you should be able to write the equation.

Example 10

Carbon dioxide and water are two of the three products of the reaction between magnesium carbonate and hydrochloric acid. Write the equation.

You have two reactants and two out of three products identified. See if you can write the formula of the third product, too, as you complete Step 1 of the procedure. If you get that far, balancing is easy. Go all the way.

. . . .

10a. $MgCO_3 + 2\ HCl \rightarrow MgCl_2 + CO_2 + H_2O$

The reactant formulas suggest a typical double displacement equation, which probably describes accurately what first occurs in the reaction. The H_2CO_3 you would expect as a product is unstable, however, and decomposes into CO_2 and H_2O. Even without this reasoning, you could conclude that $MgCl_2$ is the unnamed product. The only reacting ions not accounted for in the identified products, CO_2 and H_2O, are Mg^{2+} and Cl^-. It is logical that $MgCl_2$ be the third product.

Experiment 11

Report Sheet

NAME ....

DATE .... SECTION ....

EQUATION-WRITING EXERCISE

Write the chemical equation for each reaction described below.

A) Combination Reactions

1) Diphosphorus trioxide is formed by direct combination of its elements.

2) Ammonia and sulfuric acid combine to form ammonium sulfate.

B) Decomposition Reactions

3) Ammonium nitrite decomposes into nitrogen and water.

4) When heated, potassium chlorate decomposes into oxygen and potassium chloride.

C) Complete Oxidation or Burning of Organic Compounds

5) Propane, C_3H_8, burns in air.

6) Acetaldehyde, CH_3CHO, is completely oxidized.

D) Oxidation – Reduction Reactions

7) Hydrogen is released when aluminum reacts with hydrochloric acid.

8) Magnesium reacts with silver nitrate solution.

E) Ion Combination Reactions

9) Barium carbonate precipitates from the reaction of barium chloride and sodium carbonate solutions.

10) Sulfuric acid neutralizes calcium hydroxide.

11) Sodium iodate and silver nitrate solutions are combined.

12) Potassium fluoride reacts with hydrobromic acid.

13) Zinc hydroxide reacts with hydrochloric acid.

Experiment 11

Report Sheet

NAME ....

DATE .... SECTION ....

F) Other Reactions

14) Copper(II) chloride and water result from the reaction of copper(II) oxide and hydrochloric acid.

15) Carbon dioxide and water are two of the three products from the reaction of sulfuric acid with sodium hydrogen carbonate.

G) Mixed Reactions

16) Hydrobromic acid reacts with potassium hydroxide.

17) Aluminum reacts with phosphoric acid.

18) Silver nitrate reacts with hydrosulfuric acid.

19) Phosphorus triiodide is formed from its elements.

20) Iron(II) chloride reacts with sodium phosphate.

21) Sugar, $C_{12}H_{22}O_{11}$, is burned in air.

22) Sugar, $C_{12}H_{22}O_{11}$, breaks down to carbon and water when heated.

23) Lithium hydroxide solution comes from the reaction of lithium oxide and water.

24) Magnesium sulfate reacts with sodium hydroxide.

25) Chlorine reacts with a solution of sodium iodide.

26) Nickel hydroxide reacts with sulfuric acid.

Experiment 11
Report Sheet

NAME ....

DATE .... SECTION ....

27) Barium peroxide, BaO_2, decomposes into oxygen and barium oxide.

28) Ammonia is formed from its elements.

29) Butyl alcohol, C_4H_9OH, is oxidized completely.

30) Water is driven from copper sulfate pentahydrate, $CuSO_4 \cdot 5\ H_2O$, with heat.

31) Magnesium nitride is formed by its elements.

32) Sulfuric acid reacts with potassium nitrite.

NAME ....

Experiment 11

Advance Study Assignment

DATE .... SECTION ....

1) What is the first step in writing a chemical equation?

2) What two things must *not* be done when balancing an equation?

3) What reactant is frequently not named in a burning reaction? What are the products of all complete oxidations of compounds consisting of carbon, hydrogen, and possibly oxygen?

4) Name two kinds of chemical reactions that are described by double displacement equations.

5) What term is used to describe the equation that is written for a redox reaction in which an element appears to take the place of another element in a compound?

Experiment 12Ⓜ

Mole Ratio for a Chemical Reaction

Performance Goals

12–1 Carry out a reaction between a measured amount of sodium hydrogen carbonate, $NaHCO_3$, and hydrochloric acid, HCl.

12–2 By weighing the solid product, derive the mole ratio in the reaction between the reactant and the product.

CHEMICAL OVERVIEW

In this experiment you will carry out a reaction where one of the products is a gas and the other is a solid.

$$NaHCO_3(s) + HCl(aq) \rightarrow NaCl(aq) + H_2O + CO_2(g) \qquad (12.1)$$

The above process will be carried out in a test tube. The "container" (see page 8) for this experiment, however, will be a beaker and the test tube. The beaker is necessary to hold the test tube on the balance in an upright position. Be sure to use the same beaker for each weighing.

To make sure that all the liquid has been evaporated, the solid residue should be "heated to constant mass." After the first heating, cooling, and weighing of the product, the test tube is heated, cooled and weighed again. The two weighings should be within the plus-and-minus uncertainty of the balance used. If this is not the case, the heating procedure should be repeated again until the difference in mass is within the limits of the balance (about 0.005 for a milligram balance) between two consecutive weighings.

SAFETY PRECAUTIONS

Hydrochloric acid is corrosive. Avoid breathing the vapors and contact with skin. If acid should spill on you, *immediately* wash it off with *plenty* of water. *Extreme caution* should be exercised when heating the $NaHCO_3$ with HCl. Vigorous bubbling will occur, and when heating the contents of the test tube it will have a tendency to "shoot out." Be sure to move the test tube slowly in and out of the flame, while cautiously shaking the test tube. Always keep the test tube in a slanted position to expose the maximum area of the contents to heating. *Do not* point the open end of the test tube at anyone (including yourself!).

PROCEDURE

Note: Use a milligram balance for all measurements. Record data to the nearest 0.001g.

A) Place a clean and dry 18 × 150 mm test tube in a beaker. Weigh this "container" and records its mass on the report sheet.

B) Add about 0.25 – 0.35 g of sodium hydrogen carbonate, $NaHCO_3$, to the test tube. Weigh the beaker, test tube, and contents. Record the mass.

C) One drop at a time, cautiously add 3 M hydrochloric acid, HCl, to the test tube. After each addition, gently shake the test tube until the reaction stops. Continue adding acid until there is no evidence of any further reaction (no more bubbles form). At this point no more solid should be present.

D) Evaporate all fluid by slowly moving the test tube in and out of the flame. Remember to keep the test tube slanted and shake it gently. Proceed until all liquid has evaporated.

E) Remove the test tube from the flame and test for water vapor by inverting a clean and dry test tube over the mouth of your test tube. If you see condensation, heat the test tube for an additional 5 minutes and test again. Alternatively, after the contents of the test tube appear dry, place the test tube in a 110 – 120°C oven and heat for 20 minutes.

F) Cool the test tube to room temperature. Place the test tube in the beaker used for earlier weighing and determine the mass.

CALCULATIONS

Using only numbers from your measurements, calculate by difference the initial mass of the sodium chloride. From these quantities, using the appropriate molar masses, calculate the number of moles of each compound.

$$\text{Moles} = \frac{\text{mass (g)}}{\text{molar mass (g/mole)}} \tag{12.2}$$

Calculate the mole ratio of $NaHCO_3$ to NaCl. Compare this value to the theoretical value. Calculate the percent error.

Experiment 12
Work Page

NAME ....

DATE .... SECTION ....

MASS DATA

Mass of container (g)	
Mass of container + $NaHCO_3$ (g)	
Mass of container + NaCl (g) 1st heating	
Mass of container + NaCl (g) 2nd heating	

RESULTS

Mass of $NaHCO_3$ (g)	
Mass of NaCl (g)	

CALCULATIONS

A) Moles of $NaHCO_3$ and NaCl:

B) Mole ratio:

C) How does your experimental ratio compare to the theoretical? If they are different, explain what might have caused the difference.

D) Calculate the percent error.

Experiment 12

Report Sheet

NAME ....

DATE .... SECTION ....

MASS DATA

Mass of container (g)	
Mass of container + $NaHCO_3$ (g)	
Mass of container + NaCl (g) 1st heating	
Mass of container + NaCl (g) 2nd heating	

RESULTS

Mass of $NaHCO_3$ (g)	
Mass of NaCl (g)	

CALCULATIONS

A) Moles of $NaHCO_3$ and NaCl:

B) Mole ratio:

C) How does your experimental ratio compare to the theoretical? If they are different, explain what might have caused the difference.

D) Calculate the percent error.

NAME ....

Experiment 12
Advance Study Assignment

DATE .... SECTION ....

1) In the laboratory a student carried out the reaction

$$CaCO_3 + 2HCl \rightarrow CaCl_2 + H_2O + CO_2$$

The following data were collected:

Mass of container	48.365g
Mass of container + $CaCO_3$	48.638g
Mass of container + contents after evaporation of the liquid	48.664g

Based on the data above, calculate:

A) Starting mass of $CaCO_3$

B) Mass of $CaCl_2$ obtained

C) Mole ratio of $CaCO_3$ to $CaCl_2$

2) List possible sources of error in this experiment.

Experiment 13Ⓜ

Types of Chemical Reactions

Performance Goals

13–1 Carry out various chemical reactions.
13–2 Demonstrate that during chemical reactions mass is conserved.

CHEMICAL OVERVIEW

Chemical reactions can be classified as:

a) Combination or synthesis reactions in which two or more substances combine to form a single product.
b) Decomposition reactions, which are the opposite of the combination reactions, in that a compound breaks down into simpler substances.
c) Complete oxidation (burning) of organic compounds. In these reactions an organic compound reacts with oxygen yielding carbon dioxide, $CO_2(g)$ and water, $H_2O(g)$ or $H_2O(l)$.
d) Precipitation reactions, when the cation from one compound reacts with the anion of another compound yielding a solid product (precipitate). These reactions are also called double replacement or ion combination reactions since ions of the two reactants appear to change partners.
e) Oxidation-reduction reactions, during which one of the reactants gives off electrons (gets oxidized) and the other gains electrons (gets reduced).
f) Acid-base reactions, also called neutralization reactions, in which an acid reacts with a base yielding a salt and (usually) water.

In this experiment, we will carry out several different reactions, starting and ending with metallic copper.

SAFETY PRECAUTIONS

In this experiment you will use fairly concentrated acids and bases. When in contact with skin, most of these chemicals cause severe burns if not removed promptly. Always wear goggles when working with these chemicals. Reacting metal with nitric acid should *only* be carried out *in the hood.* Be careful when using a boiling water bath. Replenish the water from time to time as it becomes necessary.

PROCEDURE

1. DISSOLUTION OF COPPER

A) Weigh a clean and dry 25-mL Erlenmeyer flask on a milligram balance. Record this value on the report sheet.

B) Place about 100 mg of metallic copper (wire or granules) into the flask. Weigh the metal and flask to the milligram and record the mass.

C) *In the hood*, add 2 mL of 6 M nitric acid, HNO_3, to the flask and warm the contents on a hot plate. Brown vapors will form as the metal dissolves. Continue heating until no more brown fumes exist over the solution. Allow the solution to cool to room temperature, then add 2 mL of deionized water.

2. PREPARATION OF COPPER(II) HYDROXIDE

A) To the solution prepared above, carefully add 6 M sodium hydroxide, NaOH, drop by drop, until the solution is basic to red litmus paper (red paper turns blue). You can use magnetic stirring or swirl the contents of the flask while adding the NaOH. A light blue precipitate, $Cu(OH)_2$, will form.

3. PREPARATION OF COPPER(II) OXIDE

A) While stirring, heat the flask and its contents in a boiling water bath or on a magnetic-stirring hot plate. In about 5 minutes the blue $Cu(OH)_2$ will be converted to the black copper(II) oxide.

B) Allow the mixture to cool to room temperature. Remove the magnetic stirrer, if used, using forceps. Rinse with a small amount of deionized water, collecting the rinse in the Erlenmeyer flask.

C) Set up a vacuum filtration apparatus using a Hirsch or small Buchner funnel. Transfer the black precipitate into the funnel, rinse the flask with 1 to 2 mL of deionized water. The filtration may be a bit slow towards the end, due to the small particles of copper(II) oxide plugging up the filter paper pores. Wash the precipitate with 1 to 2 mL of deionized water. Discard the filtrate.

4. CONVERTING COPPER(II) OXIDE TO COPPER(II) CHLORIDE

A) Pour 6 mL of 6 M hydrochloric acid, HCl, into a 50-mL beaker. Using a spatula, transfer the black precipitate and the filter paper to the acid solution. Stir the mixture with a glass stirring rod until the precipitate is completely dissolved. If needed, heat the solution on a hot plate. Remove the filter paper and rinse it with 1 to 2 mL of deionized water, adding the rinse to the green solution. Do not use metal forceps or tweezers since they will contaminate the solution. If some precipitate is stuck on the funnel, hold it over the beaker and rinse it with 1 to 2 mL of 6 M hydrochloric acid solution. Rinse the funnel with 1 to 2 mL of deionized water. This rinse should also be collected in the beaker.

5. RECOVERING THE METALLIC COPPER

A) Weigh about 200 mg of zinc powder on a piece of pre-weighed weighing paper.

B) *In the hood,* ***very*** *slowly* add small portions of the zinc powder to the copper(II) chloride solution. Stir after each addition. You will observe the formation of copper metal and vigorous evolution of hydrogen gas.

C) Test for completeness of the reaction by adding 2 to 3 drops of your solution to 1 mL of concentrated ammonia, NH_3, in a small test tube. If a blue color appears, the reaction is not yet complete. Add a few more *small* portions of zinc powder and test again.

D) After the reaction is complete, add 5 mL of 3 M hydrochloric acid to the solution in the beaker and stir with a glass rod. This will hasten the removal of excess zinc in your mixture. Metallic copper does not react with hydrochloric acid. Allow the solution to stand for 5 minutes, stirring occasionally.

E) Using a small funnel and a pre-weighed filter paper, first pour the solution into the funnel, then transfer the solid copper into the funnel. Use deionized water to rinse the beaker and be sure all solid has been collected in the funnel. Wash the copper twice with 2 mL portions of deionized water.

F) Remove the filter paper and copper from the funnel, spread it out on a watch-glass, and allow it to air-dry. At the beginning of the next lab period weigh the copper to the milligram and record the mass on the report sheet. Calculate the percentage of recovery.

Experiment 13

Work Page

NAME ....

DATE .... SECTION ....

MASS DATA

Mass of flask (g)	
Mass of flask + Cu (g)	
Mass of weighing paper (g)	
Mass of weighing paper + Zn (g)	
Mass of filter paper (g)	
Mass of filter paper + Cu (g)	

RESULTS

Mass of Cu, initial (g)	
Mass of Zn (g)	
Mass of Cu, recovered	
Percent recovery	

Show your calculation, for the percentage of recovery, below:

Classify each reaction as synthesis, decomposition, precipitation, neutralization or oxidation-reduction:

Part 1) ____________________

2) ____________________

3) ____________________

4) ____________________

5) ____________________

NAME ....

DATE .... SECTION ....

Experiment 13

Report Sheet

MASS DATA

Mass of flask (g)	
Mass of flask + Cu (g)	
Mass of weighing paper (g)	
Mass of weighing paper + Zn (g)	
Mass of filter paper (g)	
Mass of filter paper + Cu (g)	

RESULTS

Mass of Cu, initial (g)	
Mass of Zn (g)	
Mass of Cu, recovered	
Percent recovery	

Show your calculation, for the percentage of recovery, below:

Classify each reaction as synthesis, decomposition, precipitation, neutralization or oxidation-reduction:

Part 1) ______

2) ______

3) ______

4) ______

5) ______

Experiment 13

Advance Study Assignment

NAME ....

DATE .... SECTION ....

1) Balance each of the following equations and classify the reactions:

   a) $Cu(s) + HNO_3(aq) \rightarrow Cu(NO_3)_2(aq) + H_2O + NO_2(g)$

   b) $Cu(NO_3)_2(aq) + NaOH(aq) \rightarrow Cu(OH)_2(s) + NaNO_3(aq)$

   c) $Cu(OH)_2(s) \rightarrow CuO(s) + H_2O$

   d) $CuO(s) + HCl(aq) \rightarrow CuCl_2(aq) + H_2O$

   e) $CuCl_2(aq) + Zn(s) \rightarrow Cu(s) + ZnCl_2(aq)$

2) Define an oxidation-reduction reaction

3) If you started with 0.108 g of copper and at the end of the experiment you recovered 0.099 g; calculate the percent recovery.

Experiment 14Ⓜ

Qualitative Analysis of Some Common Ions

Performance Goals

14–1 Conduct tests to confirm the presence of known ions in a solution.

14–2 Analyze an unknown solution for certain ions.

CHEMICAL OVERVIEW

When we analyze an unknown solution, two questions come to mind: (a) what ions are present in the solution, and (b) what is their concentration? The first question can be answered by performing a **qualitative analysis,** and the second by a **quantitative analysis.** These two broad categories are known collectively as **analytical chemistry.** In this experiment, you will perform a qualitative analysis.

The general approach to finding out what ions are in a solution is to test for the presence of each possible component by adding a reagent that will cause that component, if present, to react in a certain way. This method involves a series of tests, one for each component, carried out on separate samples of solution. Difficulty sometimes arises, particularly in complex mixtures, because one of the species may interfere with the analytical test for another. Although interferences are common, many ions in mixtures can usually be identified by simple tests.

In this experiment, you will analyze an unknown mixture that may contain one or more of the following ions in solution:

$$CO_3^{2-} \quad Cl^- \quad SCN^- \quad SO_4^{2-}$$

$$PO_4^{3-} \quad Cu^{2+} \quad Al^{3+} \quad Fe^{3+}$$

First, you will perform the various tests designed to detect the presence of individual ions. Once you have observed these specific reactions, you will obtain the unknown solution from your instructor. Then, taking small portions of this solution, you will run each reaction again to determine which ions are present and which are absent.

This experiment is designed to test the behavior of only a few ions. More complex schemes are used for a more complete qualitative analysis.

SAFETY PRECAUTIONS AND DISPOSAL METHODS

In some tests you will be required to use fairly concentrated acids and bases. When in contact with skin, most of these chemicals cause severe burns if not removed promptly. Wear goggles when working with any of the reagents required in this experiment.

Discard solutions containing heavy metal precipitates in the container provided.

PROCEDURE

A boiling water bath is required for some of the tests you are to perform. Pour about 100 mL of deionized water into a 150-mL beaker and heat it to boiling. Maintain it at that temperature throughout the experiment, replenishing the water from time to time as it becomes necessary.

1. TEST FOR THE CARBONATE ION, CO_3^{2-}

Cautiously add about 10 drops of 1 M HCl to 10 drops of 1 M Na_2CO_3 in a small-size test tube. Bubbles of a colorless and odorless gas, carbon dioxide, usually appear immediately in the presence of the carbonate ion. If the bubbles are not readily apparent, warm the solution in the hot-water bath and stir.

2. TEST FOR THE SULFATE ION, SO_4^{2-}

Cautiously add about 10 drops of 1 M HCl to 10 drops of 0.5 M Na_2SO_4. Then add 3 to 4 drops of 1 M $BaCl_2$. A white, powdery precipitate of $BaSO_4$ indicates the presence of SO_4^{2-} ions in the sample.

3. TEST FOR PHOSPHATE ION, PO_4^{3-}

Add 1 M HNO_3 to 10 drops of 0.5 M Na_3PO_4 until the solution is acidic. (Test by dipping a stirring rod into the solution and touching the wet rod to a strip of blue litmus paper. The solution is acidic if the color changes to red.) Then add 5 drops of 0.5 M $(NH_4)_2MoO_4$ and heat the test tube in a hot-water bath. A powdery, light yellow precipitate indicates the presence of PO_4^{3-} ions.

Caution: The molybdate solution is yellow. Be sure you see a precipitate before concluding that PO432 is present.

4. TEST FOR THIOCYANATE ION, SCN^-

Add 10 drops of 3 M $HC_2H_3O_2$ to about 10 drops of 0.5 M KSCN and stir with a glass rod. Add 1 or 2 drops of 0.1 M $Fe(NO_3)_3$. A **deep** red color formation is proof of the presence of SCN^- ions.

5. TEST FOR CHLORIDE ION, Cl^-

A) Add 1 mL of 1 M HNO_3 to about 1 mL of 0.5 M NaCl. Add 2 or 3 drops of 0.1 M $AgNO_3$. A white precipitate of AgCl confirms the presence of chloride ion.

B) If thiocyanate ion is present, it will interfere with this test, since it also forms a white precipitate with $AgNO_3$. If the sample contains SCN^- ion, put 10 drops of the solution in a medium-size test tube and add 10 drops of 1 M HNO_3. Boil the solution gently until the volume is reduced to half. This procedure will oxidize the thiocyanate and remove the interference. Then perform the chloride ion test as previously explained.

6. TEST FOR ALUMINUM ION, Al^{3+}

Add 1 M NH_3 dropwise to about 10 drops of 0.5 M $AlCl_3$ until the solution is basic (red litmus turns blue). A white, gelatinous precipitate, $Al(OH)_3$, will form. Add 3 M $HC_2H_3O_2$ dropwise until the solid dissolves. Stir and add 2 drops of cathecol violet reagent. A blue solution indicates the presence of Al^{3+} ions.

7. TEST FOR THE COPPER(II) ION, Cu^{2+}

Add concentrated ammonia drop by drop to about 10 drops of 0.5 M $CuSO_4$. The development of a dark blue color is proof of the presence of copper(II) ion.

8. TEST FOR THE IRON(III) ION, Fe^{3+}

This test is essentially the same as the test used for the thiocyanate, SCN^-, ion. To about 10 drops of 0.1 M $Fe(NO_3)_3$ add 3 or 4 drops of 0.5 M KSCN. A **deep** red color will appear if Fe^{3+} ions are present.

9. ANALYSIS OF AN UNKNOWN

When you have completed all of the tests, obtain an unknown from the instructor. Analyze it by using 10 drop portions of the unknown and then applying the tests to separate portions. The unknown will contain three to five of the ions on the list, so your test for a given ion may be affected by the presence of others. If a test does not go quite as expected, try to figure out why the sample may have behaved as it did. When you think you have analyzed your unknown properly, you may, if you wish, make a "known" that has the same ions you have found and test it to see if it has the properties of the unknown.

Experiment 14

Work Page

NAME ....

DATE .... SECTION ....

Ions Tested	*Observations (Known)*	*Unknown*	
		Yes	*No*
CO_3^{2-}			
SO_4^{2-}			
PO_4^{3-}			
SCN^-			
Cl^-			
Al^{3+}			
Cu^{2+}			
Fe^{3+}			

Unknown Number _________ . Ions present ____

Experiment 14

Report Sheet

NAME ....

DATE .... SECTION ....

Ions Tested	*Observations (Known)*	*Unknown*	
		Yes	*No*
CO_3^{2-}			
SO_4^{2-}			
PO_4^{3-}			
SCN^-			
Cl^-			
Al^{3+}			
Cu^{2+}			
Fe^{3+}			

Unknown Number _________ . Ions present ____

Experiment 14

Advance Study Assignment

NAME ....

DATE .... SECTION ....

1) An unknown that might contain any of the eight ions studied in this experiment (but no other ions) has the following properties:

   a. On addition of 1 M HCl, bubbles form.
   b. When 0.1 M $BaCl_2$ is added to the acidified unknown, a white precipitate results.
   c. When 0.1 M $AgNO_3$ is added to the unknown, a clear solution results.

   On the basis of the preceding information, classify each of the following ions as present (P), absent (A), or undetermined (U) by the tests described:

   CO_3^{2-} ______; SO_4^{2-} ______; PO_4^{3-} ______; SCN^- ______;

   Cl^- ______; Al^{3+} ______; Cu^{2+} ______; Fe^{3+} ______.

OPTIONAL ASSIGNMENT

Write net ionic equations for the reactions in this experiment in which the following ions are detected:

a) CO_3^{2-}: ____

b) SO_4^{2-}: ____

c) Cl^-: ____

Experiment 15

Molecular Models: A Study Assignment

Performance Goals

15–1 Write Lewis (electron dot) diagrams for molecules and ions formed by representative elements.
15–2 Predict the polarity of bonds and molecules formed by representative elements.
15–3 Predict bond angles and shapes of molecules and polyatomic ions.
15–4 Construct models for some covalently bonded species.

CHEMICAL OVERVIEW

Chemical bonds are the forces that hold atoms together in a compound. In this experiment we will study only covalently bonded species. A covalent bond is formed when a pair of electrons is shared by two atoms. Bonds in which the electrons are shared equally by the two nuclei are described as **nonpolar.** We expect to find nonpolar bonds whenever two identical atoms are joined, such as H_2 or Cl_2.

When different atoms are joined together, the polarity of the bond depends on the electronegativity difference between the two elements. In the HF molecule, for example, the electron density is greater around the fluorine atom and the bond has a nonsymmetrical or nonuniform electron distribution. This type of a bond is referred to as **polar.**

We find experimentally that bonds formed between atoms that differ in electronegativity by 0.4 unit or less (such as the C—H bond) behave very much like pure nonpolar bonds and therefore may be classified as "essentially nonpolar." Bonds with an electronegativity difference greater than 1.7 are regarded as **ionic** bonds. (HF is an exception.) Electronegativity values for some common elements are listed in Table 15–1.

Table 15–1 Selected Electronegativity Values

Element	*Electroneg.*	*Element*	*Electroneg.*
H	2.1	Si	1.8
B	2.0	P	2.1
C	2.5	S	2.5
N	3.0	Cl	3.0
O	3.5	Br	2.8
F	4.0	I	2.5

When determining whether a *molecule* is polar or not, its structure as well as the type of bonds in the molecule must be considered. If the molecule *has* polar bonds *and* is *nonsymmetrical,* it will be polar. On the other hand, even if the molecule has polar bonds, but is structurally symmetrical, the species will be nonpolar.* Obviously, if there are no polar bonds in a molecule, it will be nonpolar.

* The question of polarity does not apply to ionic species since they carry an overall charge.

In this experiment you will be asked to determine the polarity of certain bonds and, after considering the geometry of the molecule, decide whether it is polar or nonpolar.

Lewis (Electron Dot) Diagrams

In most stable molecules or polyatomic ions, each atom tends to acquire a noble-gas structure by sharing electrons. This tendency is often referred to as the **octet rule.** One way to show the structure of an atom or a molecule is using dots to represent the outermost s and p electrons (the so-called **valence electrons**). For the A group elements the number of valence electrons is the same as the group number in the periodic table.

Group	IA	IIA	IIIA	IVA	VA	VIA	VIIA	0
No. of valence electrons	1	2	3	4	5	6	7	8
Lewis structure	Li·	B̈e	:Ḃ	:Ċ·	:Ṇ̇·	:Ö̤·	:F̤̈·	:N̤̈e:

In writing Lewis diagrams we usually do not attempt to show which atom the valence electrons come from; we simply indicate a shared pair of electrons by either two dots or a straight line connecting the atoms. Unshared pairs, also called lone pairs, are indicated by dots written around the elemental symbols. Example:

H_2O 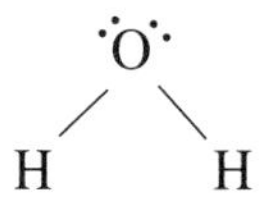 NH_3 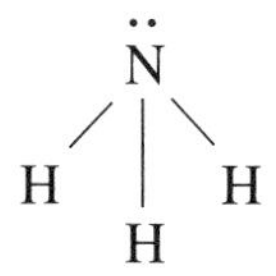

Atoms in polyatomic ions are held together by covalent bonds. In the ions we will consider in this assignment, all atoms have a noble-gas structure.

Occasionally there are too few electrons available in a species to allow an octet to exist around each atom with only single bonds. In these instances multiple (double or triple) bonds will form.

Writing Lewis Structures

The Lewis diagrams of many species can be drawn by inspection. For more complex species, however, the following procedure is helpful:

1) **Draw a tentative diagram for the molecule or ion, joining atoms by single bonds. Place electron dots around each symbol except hydrogen so the total number of electrons for each atom is eight.** In some cases, only one arrangement of atoms is possible. In others, two or more structures may be drawn. Ultimately chemical or physical evidence must be used to decide which of the possible structures is correct. A few general rules will help you in drawing diagrams that are most likely to be correct:
   a) A hydrogen atom always forms one bond; a carbon atom normally forms four bonds.
   b) When several carbon atoms appear in the same molecule, they are often bonded to each other. In some compounds they are arranged in a closed loop; however, we will avoid such so-called cyclic compounds in this assignment.
   c) In compounds or ions having two or more oxygen atoms and one atom of another nonmetal, the oxygen atoms are usually arranged *around* the central nonmetal atom.
   d) In an oxyacid (hydrogen + oxygen + a nonmetal, such as H_2SO_4 or HNO_3), hydrogen is usually bonded to an oxygen atom, which is then bonded to the nonmetal: H—O—X, where X is a nonmetal.

2) **Count the electrons in your diagram.**
3) **Find the total number of valence electrons available.** For a molecule, this is the sum of the valence electrons contributed by each atom in the molecule. For a polyatomic ion, this total must be adjusted to account for the charge on the ion. An ion with a -1 charge will have one more electron than the number of valence electrons in the neutral atoms; a -2 ion, two more; a $+1$ ion, one less; and so forth.
4) **Compare the numbers in Steps 2 and 3. If they are the same, the diagram is complete. If they are different, modify the diagram with multiple bonds.** If your diagram has two electrons more than the number available, there will be one double bond. If the difference is four, there will be a triple bond or two double bonds. Multiple bonds should be used only when necessary, and then as few as possible should be used.

There are some exceptions to the octet rule. It is not possible to write a Lewis diagram that has eight electrons around each atom if the total number of valence electrons is odd. Also, for compounds of Group IA, IIA, and IIIA elements, there are not enough electrons to satisfy the octet rule.

Electron Pair and Molecular Geometry

There are several theories that are used to explain the geometry (three-dimensional shape) of molecules. In this experiment we will use the valence shell electron pair repulsion (VSEPR) theory. According to this theory, electrostatic repulsion arranges the electron pairs surrounding an atom so that they are as far from each other as possible. This arrangement is the **electron pair geometry.** The molecular geometry, the arrangement of *atoms* around the central atom, is a direct result of the electron pair geometry.

Let us consider cases in which the central atom is surrounded by 2, 3, or 4 electron pairs.

1. **Two pairs of electrons.** Two pairs of electrons around a central atom are farthest from each other when they are on opposite sides of that atom. The electron pairs and the central atom are on the same straight line. The electron pair geometry is **linear.** Both electron pairs bond atoms to the central atom, so all three atoms are also on the same line. The molecular geometry is linear too.
2. **Three pairs of electrons.** Three electron pairs will be farthest apart when they are directed toward the corners of an equilateral triangle with the central atom at its center. The atom and all electron pairs are in the same plane, so the electron pair geometry is called **trigonal planar,** or **planar triangular.** If all three electron pairs are bonding pairs, the molecule is also planar triangular, with 120° bond angles. If one of the electron pairs is a lone pair, the molecular geometry is **bent** or **angular.**
3. **Four pairs of electrons.** Four pairs of electrons in three dimensions are farthest apart when located at the corners of a tetrahedron with the central atom in its center. If all four electron pairs are bonding pairs, as in CH_4, the central carbon atom is in the center of the tetrahedron and the four hydrogen atoms are at its corners. Both the electron pair and molecular geometries are **tetrahedral.** The bond angles are 109°.

   If there are three bonding pairs and one lone pair, as when nitrogen is the central atom, VSEPR still predicts a tetrahedral electron pair geometry with a tetrahedral angle for the bonding electron pairs. (In fact, the angle is slightly less than tetrahedral, but we will call it a "tetrahedral" angle.) The molecule has the shape of a low pyramid; its geometry is **trigonal pyramidal.** Two bonding pairs and two lone pairs around the central atom again yield close to a tetrahedral angle. The three-atom molecule has a **bent** geometry.

These electron pair and molecular geometries are summarized in Table 15–2.

Multiple Bonds. Bond angles in species that contain multiple bonds indicate that the electrons in the multiple bond behave as a single pair of electrons according to VSEPR. Thus a structure such as

$$-\mathrm{C}(=\ddot{\mathrm{O}}\text{:})-\ddot{\mathrm{O}}-\mathrm{H}$$

has a planar triangular geometry around the C atom with 120° bond angles.

PROCEDURE

Obtain a molecular model kit. Draw a Lewis diagram for HCl, the first item in the table on page 181. From your diagram, and using Table 15–2 as a guide, fill in all the blanks for HCl. Then build a model of an HCl molecule. Use the model to verify the geometry you predicted. If the model and your prediction do not agree, find out why. Do not proceed to H_2O until you thoroughly understand the HCl structure.

Follow the same procedure with H_2O, the next item in the table. This time you must predict a bond angle too. Your model should confirm your predictions. Again, do not proceed to NH_3 until you thoroughly understand H_2O.

Proceed in a similar manner for all species shown in the table. From a learning standpoint, it is important that you complete each species, including the model, before you proceed to the next.

Table 15–2 Electron Pair and Molecular Geometry

Total No. Electron Pairs	*Shared*	*Unshared*	*Bond Angle*	*Electron Pair Geometry*	*Molecular Geometry*	*Example*
2	2	0	180°	Linear	Linear	BeF_2
3	3	0	120°	Trig. planar	Trig. planar	BF_3
3	2	1	120°	Trig. planar	Angular (bent)	NO_2^-
4	4	0	109°	Tetrahedral	Tetrahedral	CCl_4
4	3	1	109°	Tetrahedral	Trigonal pyramidal	NH_3
4	2	2	109°	Tetrahedral	Bent	H_2O

Experiment 15

Report Sheet

NAME ....

DATE .... SECTION ....

Instructions: For each species listed below, draw the Lewis structure first, and then complete the rest of the information requested for that species. "Build" the molecule or ion with your model kit. Verify from your model the geometry you predicted. Complete each line in the table, including the model, before proceeding to the next line. If there are more than two atoms as "center," you can only deduce the geometry separately for each one. Continue this procedure for each species listed here.

Formula	*Lewis Structure*	*Electron Pair Geometry*	*Molecular Geometry*	*Bond Angle*	*Polar*	*Nonpolar*
HCl		X		X		
H_2O						
NH_3						
BCl_3						
CO_2						
IO_2^-					X	X
PO_4^{3-}					X	X

Formula	***Lewis Structure***	***Electron Pair Geometry***	***Molecular Geometry***	***Bond Angle***	***Polar***	***Nonpolar***
C_2H_6						
C_2H_4						
C_2H_2						
C_2H_5OH		Around O	Around O			
HCHO						
CH_3—O—CH_3		Around O	Around O			
HCOOH		Around C	Around C			
$HC_2H_3O_2$		Around single-bonded O				

Experiment 15

Advance Study Assignment

NAME ....

DATE .... SECTION ....

1) Describe a polar and a nonpolar bond. Give an example of each.

2) Draw Lewis structures for CS_2, H_2Se, and ClO_3^-.

3) Give the electron pair geometry and the molecular geometry of the above three species, according to VSEPR.

4) Are CS_2 and H_2Se polar or nonpolar molecules? Explain your reasoning.

Experiment 16

Boyle's Law

Performance Goals

16–1 Given the atmospheric pressure, use an open-end manometer to measure the pressure exerted by a confined gas.

16–2 State Boyle's Law and its mathematical significance.

CHEMICAL OVERVIEW

Boyle's Law states the relationship between the volume and pressure of a fixed quantity of confined gas at constant temperature. In performing this experiment, you will measure pressure and volume in a rather simple apparatus consisting of two glass tubes connected by a piece of rubber tubing, as shown in Figure 16–1. (More elaborate commercial apparatus operate under the same principle.) One glass tube is open to the atmosphere, while the other is closed by a screw clamp applied to a rubber hose just above the end of the glass. The system contains mercury. The surface of the mercury in the open tube is subject to atmospheric pressure, whereas the surface of the mercury in the closed tube is subject to the pressure applied by the gas trapped between the surface and the point where the tube is clamped shut. The volume occupied by this trapped air is the volume of the tube between these points. These are the pressures and volumes to be measured in this experiment.

In measuring the volume of the confined gas, we make the reasonable assumption that the cross-sectional area of the glass tube is constant throughout its length. The volume is then equal to the product of this area times the length of the trapped air column. Because the area is constant, the volume is proportional

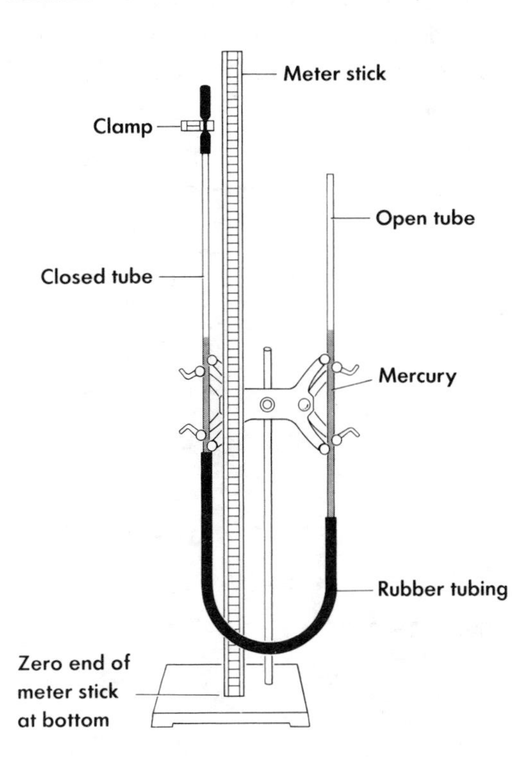

Figure 16–1. Mercury manometer.

to the length of the air space. Therefore, the length is effectively a measure of the volume. Volume will be expressed in terms of this length, measured in millimeters.

The apparatus is essentially an open-end manometer. The method of measuring pressure with such manometers is based on the equality of pressures in the two legs of the manometer *at the level of the lower liquid surface.* These pressures are the pressures exerted by the gas, or the gas and the liquid, above this level. Two possible conditions exist: (a) where the mercury level is higher in the open tube (Figure 16–2A) and (b) where the mercury level is lower in the open tube (Figure 16–2B). In the first case, Figure 16–2A, the lower liquid surface is in the closed tube. The only pressure applied at that point is the pressure of the gas, P_g. *At the same level* in the open tube, the pressure is the result of everything above that point, or $P_a + P_{Hg}$, the sum of the pressure of the atmosphere plus the pressure exerted by the mercury column. Equating these pressures at the lower liquid level, we have

$$P_g = P_a + P_{Hg} \tag{16.1}$$

P_a may be determined by reading a barometer. P_{Hg} is simply the height of the mercury column. Both pressures should be measured in millimeters of mercury (torr).

In the second case, Figure 16–2B, equating pressures at the lower liquid level gives

$$P_g + P_{Hg} = P_a \tag{16.2}$$

Solving for P_g,

$$P_g = P_a - P_{Hg} \tag{16.3}$$

If you compare Equations 16.1 and 16.3, you will observe that the pressure of the gas is always equal to the pressure of the atmosphere *plus or minus the pressure of the mercury column:*

$$P_g = P_a \pm P_{Hg} \tag{16.4}$$

From a practical standpoint, therefore, it is easier to use this generalization rather than to memorize any of the above equations, and then hope to pick out the right equation for each situation. Simply examine the physical system and decide whether the gas pressure is greater than or less than the atmospheric pressure. If it is *greater* than atmospheric, then *add* the mercury level difference to P_a; if it is *less than* atmospheric, then *subtract.*

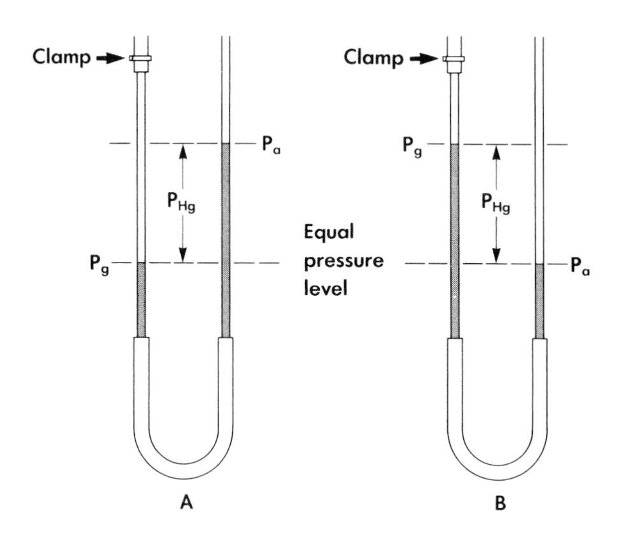

Figure 16–2.

SAFETY PRECAUTIONS

Take all possible precautions to avoid a mercury spill while performing this experiment. If this mishap does occur, notify the instructor immediately. Though apparently fun to play with, mercury must be recognized as a deadly poison. Spilled mercury must not be allowed to remain in little unseen and unreported pools in the laboratory. Furthermore, it is probably more deadly at home, where your ventilation is not as good as that in the laboratory. *No souvenirs, please!*

PROCEDURE

Note: All meter-stick readings are to be recorded in millimeters to the nearest millimeter (not in centimeters). Record atmospheric pressure to the nearest millimeter of mercury (torr).

A) Record the atmospheric pressure to the nearest torr (mm Hg) on your report sheet.

B) Be sure the clamp on the closed leg of the manometer is clamped tight *just above the end of the glass inside the rubber tube.* (Remember the assumption of a constant cross-sectional area of the glass tube as the basis of tube length being a measure of volume. If the trapped gas is in both glass and rubber tube, the assumption is not valid, and the results will be unsatisfactory.)

C) Adjust the closed leg of the manometer vertically so the top of the trapped gas is opposite the 800-, 900-, or 1000-mm mark on the meter stick. Record the meter-stick reading at that point. The closed leg is not to be moved again throughout the experiment.

D) Remove the open tube from its clamp and hold it next to the meter stick, but on the opposite side from the closed tube. Move the open tube up or down until the two mercury levels are even. Record the meter-stick reading in both the "Open" and "Closed" columns on the report sheet.

E) Raise the open tube as high as the apparatus will allow without pinching the rubber hose or causing the mercury level to rise above the top of the meter stick. Hold the tube next to the meter stick and record the mercury level in both tubes.

Note: Record these measurements just as you see them, without performing any arithmetic on them.

(When you raise the open tube, its mercury surface goes up. Which way does the mercury surface move in the closed tube?)

F) Move the apparatus as close to the edge of the desk as possible. Now lower the open tube as far as the apparatus permits, letting the rubber hose hang over the front of the desk.

Caution: Pinching the rubber hose may not be your limit this time, but rather the level of the open tube at which the mercury begins to spill out. Obviously, this must be avoided.

Again record both open and closed tube readings.

G) Select two more positions for the open tube, one between the first and second positions, and the other between the first and third. Record open and closed tube readings for each. Be sure these positions are not close to one of your earlier readings. You should end up with five fairly evenly spaced sets of readings.

At this time you will have taken all the data you need for the experiment, and the apparatus may be released to another student while you perform your calculations. As long as the next user does not open the clamp and thereby change the quantity of gas in your sample, you may always return to the apparatus for additional data or a check of one of your original settings.

CALCULATIONS

In order to calculate the pressure exerted by the confined gas, you must first find the pressure represented by the mercury level difference, P_{Hg}. This is simply the *difference* between the mercury levels in the closed and open tubes, measured directly in millimeters of mercury, or torr. Calculate this value for each set of open and closed tube readings and record it in the appropriate column in the *RESULT* table. The pressure of the gas, P_g, may now be calculated for each setting by adding P_{Hg} to or subtracting it from atmospheric pressure, P_a, as shown in the overview.

The volume of the confined gas is simply the height of the gas column, the difference between the pinch-off point and the mercury level in the closed tube, expressed and measured in millimeters.

When the pressure and volume have both been recorded, multiply them and enter the product in the $P \times V$ column.

Experiment 16
Work Page

NAME ....

DATE .... SECTION ....

DATA: FIXED VALUES

Atmospheric pressure, P_a ____________ torr

Meter-stick reading at top of trapped gas ____________ mm

Data		*Results*			
Closed Tube (mm)	*Open Tube (mm)*	P_{Hg} *(torr)*	*Pressure,* P_g *(torr)*	*Volume (mm)*	$P \times V$

Show calculations here for one line of the table:

1) When performing an experiment to determine the relationship between two variables, all other variables that might influence either of the two must be held constant. There are two other variables that are constant in your experiment. Name these.

2) Write a brief statement that describes the relationship between pressure and volume, shown by the results of your experiment. Include the conditions that must be satisfied (Question 1) in order for your summary statement to be true. Make your statement as "quantitative" as possible.

Experiment 16

Report Sheet

NAME ....

DATE .... SECTION ....

DATA: FIXED VALUES

Atmospheric pressure, P_a ____________ torr

Meter-stick reading at top of trapped gas ____________ mm

Data		*Results*			
Closed Tube (mm)	*Open Tube (mm)*	P_{Hg} *(torr)*	*Pressure,* P_g *(torr)*	*Volume (mm)*	$P \times V$

Show calculations here for one line of the table:

1) When performing an experiment to determine the relationship between two variables, all other variables that might influence either of the two must be held constant. There are two other variables that are constant in your experiment. Name these.

2) Write a brief statement that describes the relationship between pressure and volume, shown by the results of your experiment. Include the conditions that must be satisfied (Question 1) in order for your summary statement to be true. Make your statement as "quantitative" as possible.

Experiment 16

Advance Study Assignment

NAME ....

DATE .... SECTION ....

1) What variables are related by Boyle's Law?

2) If atmospheric pressure is 762 torr, calculate the pressure of the confined gas in each case that follows:

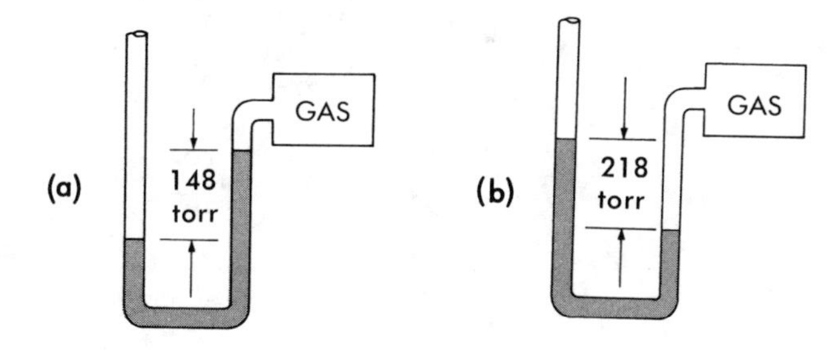

3) If P_a = 754 torr at the time the meter-stick readings shown in the illustration are taken, calculate P × V.

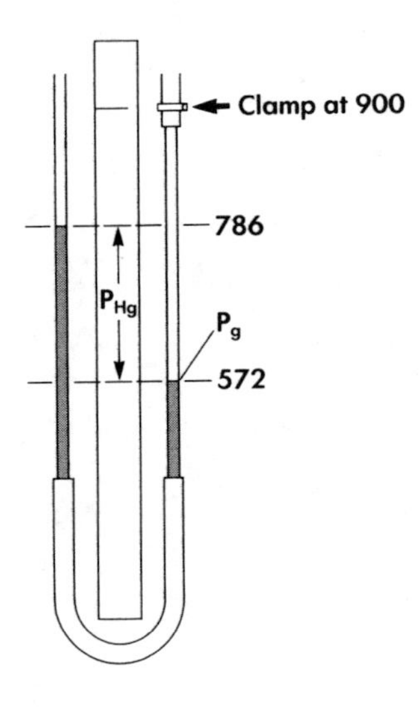

Experiment 17Ⓜ

Molar Volume of a Gas

Performance Goal

17–1 Determine the molar volume of a gas at room temperature and pressure from experimental data.

CHEMICAL OVERVIEW

The molar volume of a gas is the number of liters occupied by one mole of the gas. It is the purpose of this experiment to measure the volume occupied by a known number of moles of hydrogen. Dividing the volume in liters by the number of moles gives the molar volume in liters per mole.

The volume occupied by a fixed quantity of gas depends upon temperature and pressure. Because this entire experiment will be carried out at room temperature — the temperature at which the molar volume is to be found — this variable may be disregarded in your calculations.

The quantitative relationship between pressure and volume of a fixed amount of gas at constant temperature, known as Boyle's Law, can be expressed mathematically as

$$PV = \text{constant}, \quad \text{or} \quad P_1V_1 = P_2V_2 \tag{17.1}$$

where P is pressure, V is volume, and subscripts 1 and 2 refer to the first and second measurements of the gas sample.

In a mixture of gases, each component exerts a certain amount of pressure of its own. This pressure, called the **partial pressure,** is defined as that pressure which the component would exert if it alone occupied the total volume at the given temperature. Mathematically,

$$P = p_1 + p_2 + p_3 \cdots \tag{17.2}$$

where P is the total pressure of the mixture and $p_1, p_2, p_3, \ldots$ are the partial pressures of compounds 1, 2, 3, etc. The relationship is known as Dalton's Law of Partial Pressures.

When a confined gas is generated in contact with water, the gas becomes saturated with water vapor, producing a "wet" gas — a gaseous mixture. The partial pressure of water vapor is a function of temperature and can be found in tables. Table 17–1 contains a partial list of water vapor pressures at various temperatures.

Table 17–1 Water Vapor Pressure in Torr

Temperature (°C)	*Vapor Pressure*	*Temperature (°C)*	*Vapor Pressure*
17	14.5	23	21.1
18	15.5	24	22.4
19	16.5	25	23.8
20	17.5	26	25.2
21	18.6	27	26.7
22	19.8	28	28.3

In this experiment, hydrogen will be generated by the reaction of a measured mass of metallic magnesium with excess hydrochloric acid. The volume of gas produced will be measured in a gas measuring tube, called a *eudiometer.* Using Boyle's and Dalton's laws, the measured volume will be corrected to obtain the volume of *dry* hydrogen at room temperature. The number of moles of hydrogen will be calculated by stoichiometry from the mass of magnesium that reacted. Dividing volume by moles gives the molar volume at room temperature.

A word about significant figures: Ideally you should weigh your magnesium on an analytical balance to the nearest 0.0001 g. Most schools do not have such balances available to students in courses below the quantitative analysis level, so you are more apt to use a milligram balance that measures to the nearest 0.001 g. Ordinarily, strict application of the rules of significant figures makes this the measurement that determines the number of significant figures in your final result. It is a fact, however, that more sophisticated — and more correct — ways of handling measurement uncertainty in this experiment yield an answer with one extra digit. Therefore, if you use a milligram balance, report your final result by going one digit beyond what normal significant figure rules allow.

SAMPLE CALCULATIONS

Example 1

Calculate the pressure of dry N_2 at 26°C if it is collected over water and the total pressure is 747 torr. Water vapor pressure at this temperature is 25.2 torr.

$$P = p_{N_2} + p_{H_2O} \quad ; \quad p_{N_2} = P - p_{H_2O}$$

Substituting,

$$p_{N_2} = 747 \text{ torr} - 25 \text{ torr} = 722 \text{ torr}$$

Example 2

What volume would the dry N_2 of Example 1 occupy at 762 torr pressure if the experimentally measured volume was 80.5 mL?

Substituting into Equation 17.1,

$$P_1V_1 = P_2V_2 \quad ; \quad V_2 = \frac{P_1V_1}{P_2}$$

$$V_2 = 722 \text{ torr} \times \frac{80.5 \text{ mL}}{762 \text{ torr}} = 76.3 \text{ mL}$$

Example 3

If 0.0769 g of magnesium produced the gas in Examples 1 and 2, how many moles of hydrogen were produced? $Mg + 2\ HCl \rightarrow H_2 + MgCl_2$.

$$0.0769 \text{ g Mg} \times \frac{1 \text{ mole Mg}}{24.3 \text{ g Mg}} \times \frac{1 \text{ mole } H_2}{1 \text{ mole Mg}} = 0.00317 \text{ mole } H_2$$

Example 4

Using the results of Examples 1 to 3, find the molar volume of hydrogen at room conditions.

$$\frac{0.0763 \text{ L } H_2}{0.00317 \text{ mole } H_2} = 24.1 \text{ L } H_2\text{/mole}$$

SAFETY PRECAUTIONS AND DISPOSAL METHODS

Be careful when handling concentrated hydrochloric acid. If it comes into contact with your skin, wash it with plenty of water. Wear goggles when performing this experiment.

The hydrochloric acid solution obtained in the eudiometer can be poured down the drain. Any excess 6 M HCl should be diluted and then poured down the drain.

PROCEDURE

Note: Record the mass of magnesium in grams to the nearest 0.0001 g if you use an analytical balance, or 0.001 g if you use a milligram balance (see Overview). Record volume measurements in milliliters to the nearest 0.1 mL. Record atmospheric pressure in torr to the nearest torr.

A) Obtain a strip of magnesium that weighs between 0.072 and 0.082 g. Clean both sides by rubbing with steel wool. Weigh the magnesium, preferably on an analytical balance, and record the mass to the nearest 0.0001 g. (If you use a milligram balance, assume that calculations from this measurement are good to one digit beyond the number normally allowed by the rules of significant figures. See the Overview.)

B) Wad the magnesium into a small irregularly shaped ball. Do not fold it neatly, or pack it tightly, as the reaction requires that the acid be able to reach all surfaces of the magnesium. Wrap this ball in a small "cage" made of copper wire. Be sure that no large "holes" are present in your cage, or pieces of magnesium will escape during the reaction. Leave a 10- to 12-cm piece of wire extending from the cage (see Figure 17–1).

C) Fill a 500-mL cylinder to the top with tap water and stand it in a large beaker, tray, or the sink to catch the overflow. In selecting a place, read ahead to Step F. Make sure you can perform that step without knocking the cylinder over.

D) Carefully pour 10 mL of 6 M hydrochloric acid, HCl, into a 100-mL eudiometer, or gas measuring tube. Then, holding the tube at an angle, slowly pour deionized water down the side of the tube. When the water nears the top of the eudiometer, hold it vertically and continue to add water until the tube is full.

E) Carefully lower the copper-wire-wrapped magnesium ball into the eudiometer until it is about 5 or 6 cm into the tube, but not so deep that it extends into the calibrated portion of the eudiometer. Place a one-hole rubber stopper into the eudiometer, securing the straight piece of copper wire to the wall of the tube. Water will come out the hole in the stopper when it is placed into the eudiometer, so be ready for it.

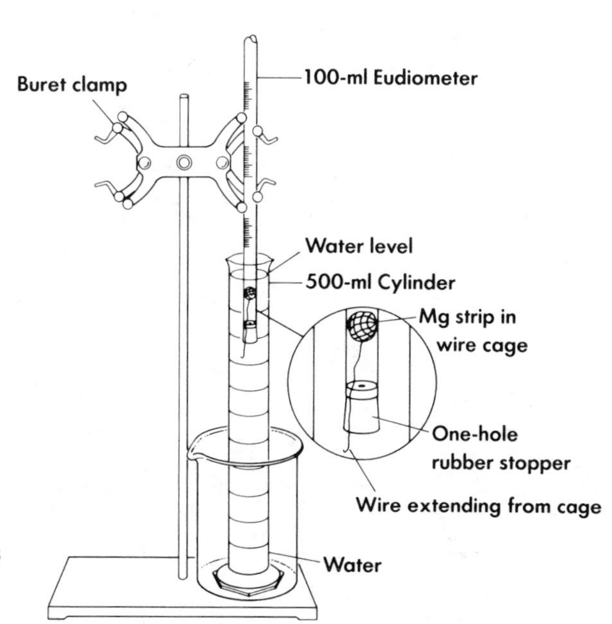

Figure 17–1. Apparatus for the volume measurement of a gas.

F) Place your finger over the hole of the stopper, invert the eudiometer and quickly lower it into the water-filled cylinder. When the stopper is beneath the surface of the water, remove your finger. Move the eudiometer and cylinder to a clamp mounted on a ring stand, as shown in Figure 17–1; a buret clamp is ideal for this purpose.

G) Allow the reaction to proceed until no more bubbles are formed.

Caution: There is apt to be more overflow of water from the cylinder, so be prepared. Tap the sides of the eudiometer gently to dislodge any trapped bubbles from the wire cage.

H) Carefully adjust the eudiometer vertically until the water level inside the tube is at the same level as the water in the cylinder. At this point the total pressure inside the tube is the same as atmospheric pressure.

I) Read and record the volume of the gas generated.

J) Read and record the atmospheric pressure in the laboratory.

K) Discard the wire cage into a wastebasket and carefully dispose of the liquid into the sink. (The liquid contains hydrochloric acid, remember?)

Experiment 17

Work Page

NAME ....

DATE .... SECTION ....

DATA

Run	*1*	*2*	*3*
Mass of Mg strip (g)			
Room pressure (torr)			
Room temperature (°C)			
Volume of H_2 gas (mL)			
Water vapor pressure (torr)			

CALCULATIONS

Moles of Mg			
Moles of H_2 gas			
Partial pressure of dry H_2 (torr)			
Volume of dry H_2 at room pressure (mL)			
Molar volume of dry H_2 at room temperature and pressure (liters/mole)			

Show setups for all calculations below for one set of data:

OPTIONAL

From *your* data obtained in this experiment, calculate the value of the universal gas constant, R. Give proper unit.

Calculate the percent error (look up the known value).

Experiment 17

Report Sheet

NAME ....

DATE .... SECTION ....

DATA

Run	*1*	*2*	*3*
Mass of Mg strip (g)			
Room pressure (torr)			
Room temperature (°C)			
Volume of H_2 gas (mL)			
Water vapor pressure (torr)			

CALCULATIONS

Moles of Mg			
Moles of H_2 gas			
Partial pressure of dry H_2 (torr)			
Volume of dry H_2 at room pressure (mL)			
Molar volume of dry H_2 at room temperature and pressure (liters/mole)			

Show setups for all calculations below for one set of data:

OPTIONAL

From *your* data obtained in this experiment, calculate the value of the universal gas constant, R. Give proper unit.

Calculate the percent error (look up the known value).

NAME ....

Experiment 17

Advance Study Assignment

DATE .... SECTION ....

1) Calculate the number of moles of gas generated when 0.165 g of aluminum reacts with hydrochloric acid.

$$2Al(s) + 6\ HCl(aq) \rightarrow 2AlCl_3(aq) + 3H_2(g)$$

2) Calculate the pressure of dry O_2 if the total pressure of O_2 generated over water is measured to be 638 torr and the temperature is 22.0°C. $p_{H_2O} = 19.8$ torr.

3) If the volume of the above O_2 sample was 56.2 mL, what volume would the dry O_2 occupy at 755 torr (assume temperature unchanged)?

Experiment 18

Molar Mass of a Volatile Liquid

Performance Goals

18–1 Determine experimentally the mass of a vapor occupying a known volume at a given temperature and pressure.

18–2 Calculate the molar mass of a volatile liquid from the mass of a given volume of vapor at a measured temperature and pressure.

CHEMICAL OVERVIEW

An "ideal gas" is one that behaves as if the molecules (a) are widely separated from each other, (b) have negligible volumes of their own compared to the space they occupy, and (c) are not subject to intermolecular attractive or repulsive forces. Under ordinary conditions in the laboratory, most gases approach the behavior of ideal gases.

In describing the behavior of gases, four variables must be considered: (a) volume, V, (b) pressure, P, (c) absolute temperature, T, and (d) amount, measured in number of moles, n. Considering the separate relationships between volume and each of the other three variables, we derive these laws:

Volume is inversely proportional to pressure at constant temperature and amount (Boyle's Law):

$$V \propto \frac{1}{P} \qquad (18.1)$$

This means, for example, that the higher the pressure on a given amount of gas (at constant T), the smaller a volume the gas will occupy.

Volume is directly proportional to absolute temperature (K) at constant pressure and amount (Charles' Law):

$$V \propto T \qquad (18.2)$$

For example, if you heat a given amount of gas at constant pressure, the volume will increase.

Volume is directly proportional to the amount of gas at constant pressure and temperature:

$$V \propto n \qquad (18.3)$$

In other words, the more gas there is (at constant T and P), the larger a volume it will occupy. Combining these relationships yields the ideal gas equation:

$$PV = nRT \qquad (18.4)$$

where R is a proportionality constant.

The units of R depend upon the units used in measuring the four variables. In working with gases in chemistry, you will measure volume in liters and temperature in degrees Celsius, which will be converted to Kelvins. If pressure is measured in atmospheres,

$$R = 0.0821 \frac{\text{(liter)(atmosphere)}}{\text{(K)(mole)}} \tag{18.5}$$

If, on the other hand, pressure is measured in mm Hg (torr),

$$R = 62.4 \frac{\text{(liter)(torr)}}{\text{(K)(mole)}} \tag{18.6}$$

The numerical value of R is the same for all gases, pure or mixtures.

A useful variation of the ideal gas equation may be derived as follows: If the mass of any chemical species, g, is divided by the molar mass, MM, the quotient is the number of moles.

$$\frac{g}{MM} = \frac{\text{grams}}{\text{grams/mole}} = \text{moles}$$

Therefore, for any pure gas, $\frac{g}{MM}$ may be substituted for its equivalent, n, in Equation 18.4:

$$PV = \frac{g}{MM} RT \tag{18.7}$$

This equation can be used to determine the molar mass of a gas or compound that is readily converted to a vapor.

In this experiment, you will place a volatile liquid into a flask of measured volume and immerse the flask in boiling water. The liquid will vaporize and drive the air out of the flask, filling the entire volume with vapor. The mass of the vapor will be measured by condensing it and weighing it. With pressure, volume, temperature, and mass all known, molar mass may be determined by direct substitution into Equation 18.7.

SAMPLE CALCULATION

Find the molar mass of a gas if a 0.895-g sample occupies 235 mL at 95°C and 1.02 atmospheres pressure.

Substitution into Equation 18.7 requires that volumes be measured in liters and temperature in Kelvins. These conversions are:

$$V = 235 \cancel{\text{mL}} \times \frac{1 \text{ liter}}{1000 \cancel{\text{mL}}} = 0.235 \text{ liter}$$

$$K = °C + 273 = 95 + 273 = 386 \text{ K}$$

Solving Equation 18.7 for molar mass and substituting the experimental values yields

$$MM = \frac{gRT}{PV} = \frac{0.895 \text{ g}}{1.02 \cancel{\text{atm}}} \times \frac{0.0821 \cancel{\text{(liter)}}\cancel{\text{(atm)}}}{\cancel{\text{(K)}}\text{(mole)}} \times \frac{368 \cancel{\text{K}}}{0.235 \cancel{\text{liter}}} = 113 \text{ g/mole}$$

In calculations with the ideal gas equation, be sure to include units. This will enable you to check whether you are setting up the problem correctly. If not, the unit of the quantity calculated will not be correct and you will know that you must recheck your setup for errors.

SAFETY PRECAUTIONS AND DISPOSAL METHODS

Be careful when handling volatile liquids. Many of these compounds can be harmful when their vapors are inhaled. Also, extreme care should be used when dealing with boiling water. If you should have to move the beaker with boiling water in it, be sure to hold the beaker securely.

Dispose of excess liquids in a stoppered bottle in the fume hood. *Do not pour liquids down the drain.*

PROCEDURE

Note: Record temperature in degrees Celsius to the nearest degree; pressure in torr to the nearest torr; mass in grams to the nearest 0.001 g; and volume in milliliters to the nearest milliliter.

A) Weigh a clean, dry 125-mL Erlenmeyer flask, along with a 5-cm × 5-cm square of aluminum foil, on a milligram balance; record the mass on the report sheet.

B) Obtain 4 to 5 mL of an unknown liquid from the instructor and place it in the flask. Record the number of the unknown. Crimp the aluminum foil fairly tightly around the neck of the flask and punch a *tiny* hole with a pin in the center of the foil.

C) Fill a 600-mL beaker about 1/3 full of water. Clamp the Erlenmeyer flask to a ring stand and immerse it in the beaker, as shown in Figure 18–1. Add more water until the flask is totally immersed, but not so deep that the water reaches the lower edge of the foil.

D) Remove the flask from the beaker and heat the water to boiling. Return the flask to its former position in the beaker and continue heating.

E) Measure and record the temperature of the boiling water and the barometric pressure.

F) As soon as *all* liquid in the flask has vaporized (no condensed beads of liquid on the walls or neck of the flask), remove the flask from the water and allow it to cool to room temperature. You may find it necessary to remove the flask from the beaker a few times to inspect it for remaining liquid, returning it promptly if liquid is present. Do not heat the flask too long, as this will lead to erroneous results.

G) As the flask cools, carefully wipe the outside, making sure that no water droplets have collected under the edge of the aluminum foil. When the flask is at room temperature, weigh the flask, foil, and condensed liquid on a milligram balance and record the mass on your report sheet.

H) To establish the volume of the gas, fill the flask completely with water and then measure its volume using a graduated cylinder. Record your reading to the nearest milliliter.

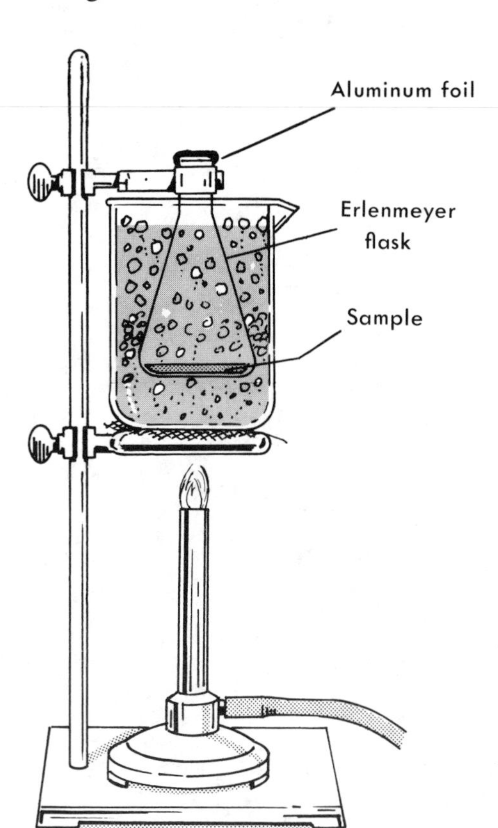

Figure 18–1. Apparatus for determining the molar mass of a volatile liquid.

CALCULATIONS

From the data, determine the volume of the gas in liters and the vapor temperature in Kelvins. (Assume the temperature is the same as that of the boiling water in which the flask was immersed.) Then find the molar mass, as shown in the sample calculation.

Experiment 18
Work Page

NAME ....

DATE .... SECTION ....

Unknown No. ______________

DATA

Run	*1*	*2*	*3*
Mass of flask and foil (g)			
Mass of flask and foil and condensed vapor (g)			
Temperature of boiling water (°C)			
Barometric pressure (torr)			
Volume of flask (mL)			

RESULTS

Mass of unknown (condensed vapor) (g)			
Volume of flask (vapor) (L)			
Temperature of vapor (K)			
Molar mass of unknown (g/mole)			

Show complete calculations for one column from the above tables:

Experiment 18

Report Sheet

NAME ....

DATE .... SECTION ....

Unknown No. ______________

DATA

Run	*1*	*2*	*3*
Mass of flask and foil (g)			
Mass of flask and foil and condensed vapor (g)			
Temperature of boiling water (°C)			
Barometric pressure (torr)			
Volume of flask (mL)			

RESULTS

Mass of unknown (condensed vapor) (g)			
Volume of flask (vapor) (L)			
Temperature of vapor (K)			
Molar mass of unknown (g/mole)			

Show complete calculations for one column from the above tables:

Experiment 18

Advance Study Assignment

NAME ....

DATE .... SECTION ....

1) How would each of the following errors affect the outcome of this experiment? Would it make the molar mass high or low? Give your reasoning in three sentences or less in each case.

   a) The hole in the aluminum foil was quite large.

   b) Water vapors condensed under the aluminum foil before the final weighing.

2) A volatile liquid was allowed to evaporate in a 43.298 g flask that has a total volume of 252 mL. The temperature of the water bath was 100°C at the atmospheric pressure of 776 torr. The mass of the flask and condensed vapor was 44.173 g. Calculate the molar mass of the liquid.

Experiment 19

Molar Mass Determination by Freezing-Point Depression

Performance Goals

19–1 Measure the freezing point of a pure substance.
19–2 Determine the freezing point of a solution by graphical methods.
19–3 Knowing the molal freezing-point constant and having determined experimentally the mass of an unknown solute, the mass of a known solvent, and the freezing-point depression, find the molar mass of the solute.

CHEMICAL OVERVIEW

When a solution is prepared by dissolving a certain amount of solute in a pure solvent, the properties of the solvent are modified by the presence of the solute. The change in such properties as the melting point, boiling point, and vapor pressure is found to be dependent on the number of solute molecules or ions in a given amount of solvent. The nature of the solute particles (molecules or ions) is not important; the governing factor is the *relative number of particles.* Properties that are dependent only on the concentration of solute particles are referred to as **colligative properties.**

If a pure liquid is cooled, the temperature will decrease until the freezing point is reached. With continued cooling, the liquid gradually freezes. As long as liquid and solid are *both* present, the temperature will remain constant. When all the liquid is converted to solid, the temperature will drop again.

A typical cooling curve for a pure solvent is shown in Figure 19–1. The dip below the freezing point is the result of supercooling, an unstable situation in which the temperature drops below the normal freezing point until crystallization begins. Supercooling may or may not occur in any single freezing process and will probably vary in successive freezings of the same sample of pure substance. As soon as crystal formation begins, the temperature rises to the normal freezing point and remains constant until all of the liquid is frozen.

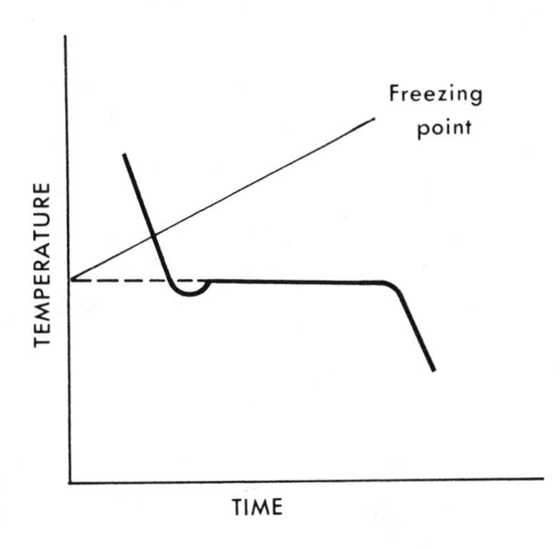

Figure 19–1. Cooling curve of a pure liquid.

The freezing point of a solution is always lower than that of the pure solvent. The difference between the freezing points of the solvent and solution is the **freezing-point depression.** This difference is proportional to the molal concentration of solute (see Equation 19.1). During the freezing of a solution, it is the solvent that freezes. Therefore, the solvent is gradually removed from the solution as the freezing progresses, leaving behind an increasingly concentrated solution. Because of this concentration increase, the freezing point drops, producing a solution freezing curve such as that in Figure 19–2. Again, the supercooling effect may or may not be observed in any single freezing experiment.

Colligative properties may be used to find the molar mass of an unknown substance. It is known, for example, that 1 mole of any molecular substance (one that does not produce ions in solution), when dissolved in 1 kg of water, lowers its freezing point by 1.86°C and raises its boiling point by 0.52°C. The freezing-point depression or boiling-point elevation for a 1.0 molal solution of a molecular substance will be the same regardless of the nature of the substance, as long as no ionization takes place. In general, the relationship between freezing-point depression and concentration of solute can be expressed as

$$\Delta T_F = K_F m \tag{19.1}$$

where ΔT_F is the freezing-point depression in °C, K_F is the molal freezing-point constant of the pure solvent, and m is the molality of the solution (moles of solute per kg of solvent).

The solvent in this experiment is naphthalene ($K_F = 6.9°C/m$). You will determine its freezing point experimentally with *your* thermometer. (The literature value of the freezing point is 80.2°C.) You will then prepare a solution from a weighed quantity of naphthalene and a weighed quantity of unknown solute, and determine experimentally its freezing point. When you substitute ΔT_F into Equation 19.1, the molality of the solution can be calculated.

Molality, by definition, is moles of solute per kilogram of solvent. This may be expressed mathematically as

$$m = \frac{\text{mole solute}}{\text{kg solvent}} = \frac{\text{g solute/MM}}{\text{kg solvent}} \tag{19.2}$$

Knowing the mass of the solute and the mass of the solvent used, the molar mass (MM) can be calculated if the molality (m) is determined experimentally.

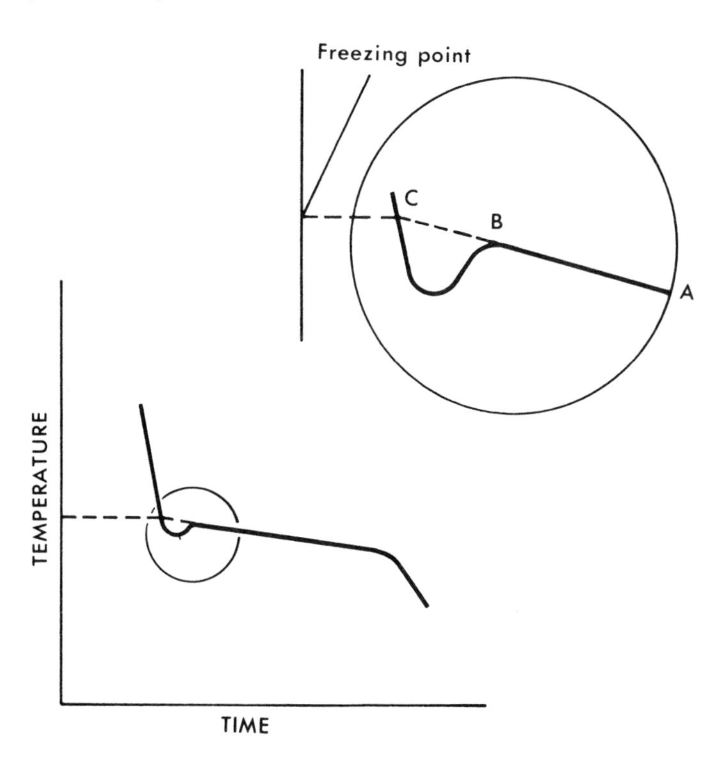

Figure 19–2. Cooling curve of a solution.

SAMPLE CALCULATIONS

Example 1

Calculate the molal concentration of a solution of 3.45 g of compound A (MM = 120 g/mole) dissolved in 50.0 g of solvent.

Converting the mass of solvent to kg and substituting the given values into Equation 19.2,

$$m = \frac{3.45 \text{ g solute}/120 \text{ g solute/mole}}{0.050 \text{ kg solvent}}$$

$$= \frac{3.45 \text{ g solute}}{0.050 \text{ kg solvent}} \times \frac{1 \text{ mole solute}}{120 \text{ g solute}} = 0.575 \text{ mole/kg solvent}$$

Example 2

Calculate the molar mass of an unknown substance if a solution containing 2.37 g of unknown in 20.0 g of water freezes at −1.52°C. K_F for water = 1.86°C/m.

The normal freezing point of water is 0.00°C, and that of the solution is −1.52°C. The freezing-point depression therefore is

$$\Delta T_F = 0.00°\text{C} - (-1.52°\text{C}) = 1.52°\text{C}$$

Solving Equation 19.1 for m and substituting,

$$m = \frac{\Delta T_F}{K_F} = \frac{1.52°\text{C}}{1.86°\text{C/m}} = 0.817 \frac{\text{mole solute}}{\text{kg solvent}}$$

Now you can calculate the molar mass using Equation 19.2:

$$0.817 \frac{\text{mole solute}}{\text{kg solvent}} = \frac{2.37 \text{ g solute/MM}}{0.0200 \text{ kg solvent}}$$

After rearranging to obtain MM,

$$MM = \frac{2.37 \text{ g solute}}{0.0200 \text{ kg solvent} \times 0.817 \dfrac{\text{mole solute}}{\text{kg solvent}}} = 145 \text{ g/mole}$$

SAFETY PRECAUTIONS AND DISPOSAL METHODS

In this experiment you will be handling various organic chemicals. Always use a spatula; do not handle crystals by hand. When using organic liquids, do not breathe their vapors.

Dispose of excess reagents as directed by your instructor. Do not use organic liquids near an open flame.

PROCEDURE

Note: Record all mass measurements in grams to the nearest 0.01 g. Record all temperature measurements in degrees Celsius to the nearest 0.1°C.

1. THE FREEZING POINT OF NAPHTHALENE

A) Assemble the apparatus shown in Figure 19–3 using a thermometer graduated in tenths of degrees Celsius. A circular stirrer may be formed from No. 18 wire. A 600-mL beaker half full of water may be used as the water bath.

B) Weigh an empty test tube to ±0.01 g on a centigram balance. Add about 10 g of naphthalene and weigh again.

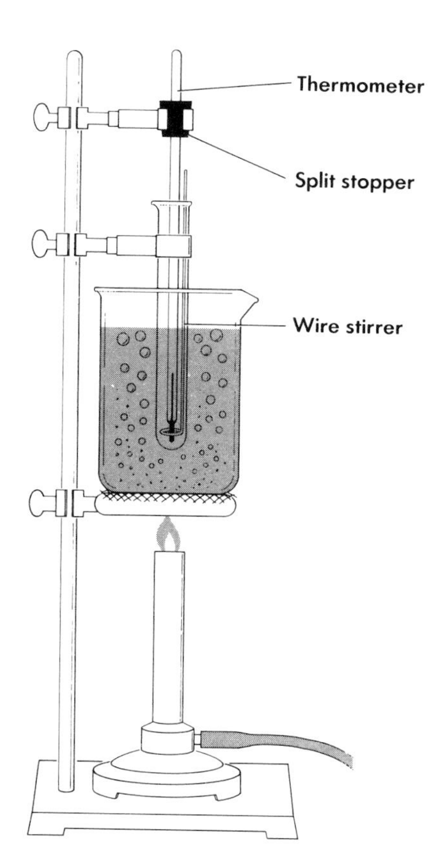

Figure 19–3. Apparatus for determining freezing point.

C) Insert the thermometer in the test tube and place the unit in the water bath. Heat the water bath until the naphthalene completely melts. Be sure the entire thermometer bulb is submerged in the molten naphthalene and is not touching the bottom or walls of the test tube.

D) Discontinue heating, remove the water bath, and allow the liquid to cool slowly. Beginning at about 90°C, record the temperature to ±0.1°C every 30 seconds for 10 minutes. Stir continuously during this period by raising and lowering the wire stirrer.

2. THE FREEZING POINT OF THE SOLUTION

A) Place a 1.0- to 1.3-g sample of unknown solute into a test tube and weigh them to the nearest 0.01 g on a centigram balance. Transfer the unknown to the test tube containing naphthalene (Procedure 1). Again weigh the empty test tube with any powder of the unknown that may not have been transferred, to determine the mass of the solute dissolved.

B) Heat the mixture in the water bath until all the naphthalene is melted and the unknown is completely dissolved. With constant stirring, determine the freezing point of the solution exactly the same way as you determined the freezing point of the pure naphthalene.

C) If time permits, remelt the solution and repeat the freezing cycle.

3. CLEAN-UP

To clean your equipment, melt the solution in the test tube and pour it onto a paper towel folded to several thicknesses. The test tube may be cleaned with hexane, acetone, or some other suitable solvent.

Caution: Do *not* pour melted naphthalene solutions into the sink. They freeze and clog drains. Many cleaning solvents are flammable and should not be used near an open flame. Clean the equipment in a hood. Do not breathe vapors.

FREEZING-POINT DETERMINATION

Estimate to $\pm 0.1°C$ the freezing point of pure naphthalene from your data. According to Figure 19–1, the freezing point of a pure substance is the constant temperature at which the substance freezes. With due allowance for supercooling, if any, this temperature should be apparent from the table. Record it on the report sheet.

To estimate the freezing temperature of the solution of the unknown in naphthalene, plot the temperature vs. time on the graph paper provided. The freezing point of the solution *of the concentration you prepared* is the point where it first begins to freeze if there is no supercooling. If supercooling does occur, the freezing point may be estimated from the graph by extrapolating the curve during freezing (from A to B in Figure 19–2) back to its intersection with the cooling line for the liquid (point C). The temperature corresponding to point C is the temperature at which freezing would have begun in the absence of supercooling — the freezing point of your solution at its initial concentration. Record the freezing temperature of the solution in the report sheet.

CALCULATIONS

From the observed freezing points of pure naphthalene and the solution, calculate the freezing-point depression. Record ΔT_F in the Results table.

Using Equation 19.1 and the molal freezing-point constant of naphthalene ($K_F = 6.9°C/m$), find the molality of your solution. Record m in the Results table.

Determine and record the kilograms of naphthalene and grams of solute in your solution.

Using Equation 19.2 and the results already determined, calculate and record the molar mass of your unknown.

Experiment 19
Work Page

NAME ....

DATE .... SECTION ....

DATA

Time–Temperature Readings (Read to $\pm 0.1°C$)

Time (Minutes)	*Naphthalene*	*Naphthalene + Unknown*			
0					
$\frac{1}{2}$					
1					
$1\frac{1}{2}$					
2					
$2\frac{1}{2}$					
3					
$3\frac{1}{2}$					
4					
$4\frac{1}{2}$					
5					
$5\frac{1}{2}$					
6					
$6\frac{1}{2}$					
7					
$7\frac{1}{2}$					
8					
$8\frac{1}{2}$					
9					
$9\frac{1}{2}$					
10					

Experiment 19
Work Page

NAME ....

DATE .... SECTION ....

Unknown No. ______________

DATA

Mass of test tube (g)			
Mass of test tube and naphthalene (g)			
Mass of test tube and unknown ________________ (g)			
Mass of test tube after transfer (g)			
Freezing point of naphthalene (°C)			
Freezing point of solution (from curve) (°C)			

RESULTS

Freezing-point depression, ΔT_F (°C)			
Molality, m (mole solute/kg solvent)			
Mass of naphthalene (g)			
Mass of unknown ________________ (g)			
Molar mass of unknown (g/mole)			

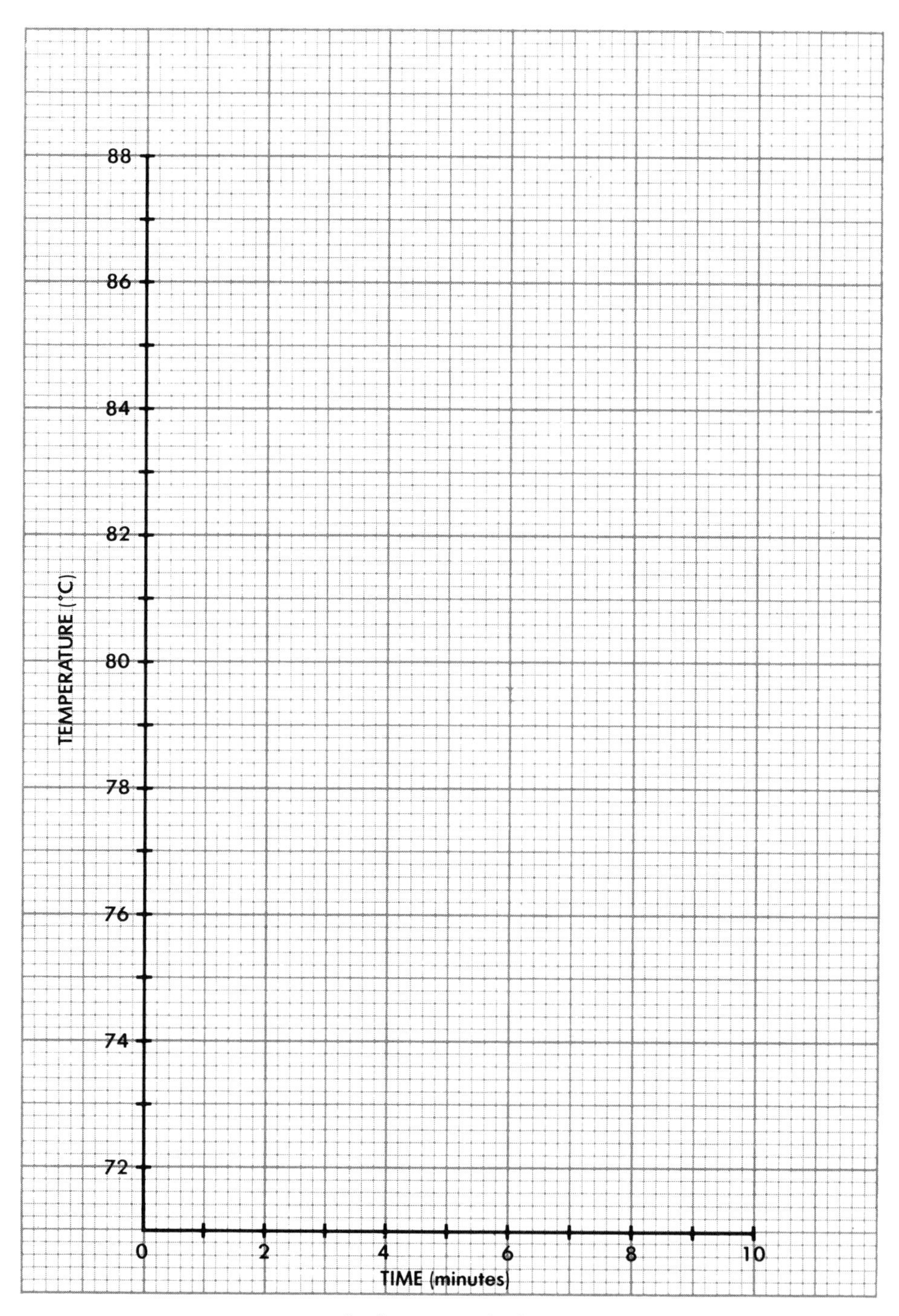

Cooling curve of solution

Experiment 19
Report Sheet

NAME ....

DATE .... SECTION ....

DATA

Time–Temperature Readings (Read to ±0.1°C)

Time (Minutes)	*Naphthalene*	*Naphthalene + Unknown*			
0					
½					
1					
1½					
2					
2½					
3					
3½					
4					
4½					
5					
5½					
6					
6½					
7					
7½					
8					
8½					
9					
9½					
10					

Experiment 19
Report Sheet

NAME ....

DATE .... SECTION ....

Unknown No. ______________

DATA

Mass of test tube (g)			
Mass of test tube and naphthalene (g)			
Mass of test tube and unknown ________________ (g)			
Mass of test tube after transfer (g)			
Freezing point of naphthalene (°C)			
Freezing point of solution (from curve) (°C)			

RESULTS

Freezing-point depression, ΔT_F (°C)			
Molality, m (mole solute/kg solvent)			
Mass of naphthalene (g)			
Mass of unknown ________________ (g)			
Molar mass of unknown (g/mole)			

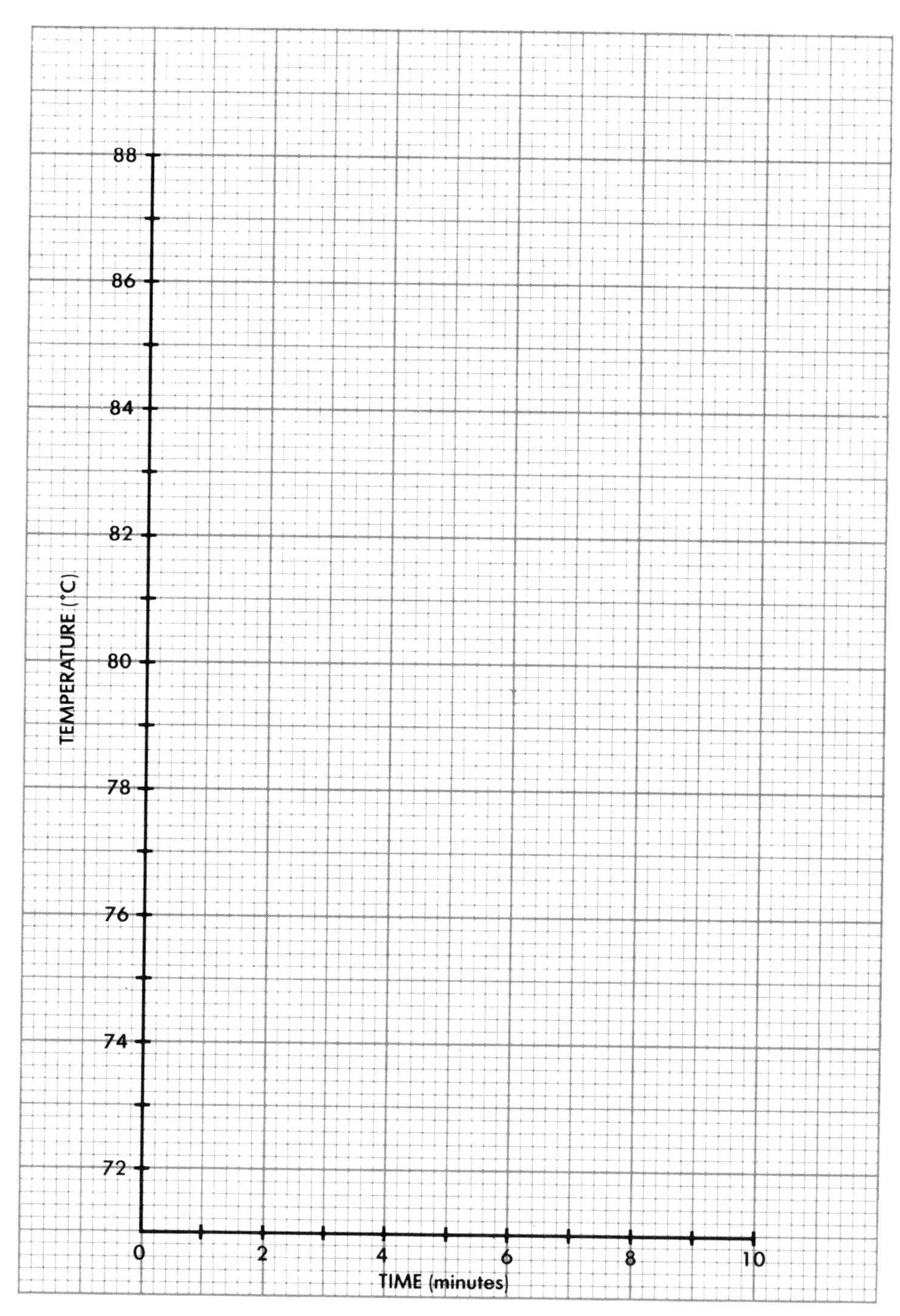

Cooling curve of solution

NAME ....

Experiment 19

Advance Study Assignment

DATE .... SECTION ....

1) Calculate the freezing-point depression for a solution prepared by dissolving 25.0 g of C_8H_{18} in 175 g of benzene. (K_F = 4.90°C/m for benzene.)

2) If 2.6 g of unknown solute are dissolved in 28 g of water, the freezing-point depression is 2.7°C.

   a) Calculate the molality of the solution if K_F = 1.86°C/m for water.

   b) Calculate the molar mass of the solute.

Experiment 20

The Conductivity of Solutions: A Demonstration

Performance Goals

20–1 Describe how the conductivity of a solution may be tested.

20–2 Basing your decision on conductivity observations, classify substances as strong electrolytes, weak electrolytes, or nonelectrolytes.

20–3 Explain the presence or absence of conductivity in an aqueous solution.

CHEMICAL OVERVIEW

Solutions of certain substances are conductors of electricity. The conductance is due to the presence of charged species (ions) that are free to move through the solution. If two metal strips, called *electrodes,* are connected to a current source (such as a battery or regular wall plug) and then immersed into a conducting solution, current will flow through the system. Ions will be attracted to the oppositely charged electrodes; i.e., positively charged ions, called *cations,* will flow to the negatively charged electrode (the *cathode*), and negatively charged *anions* will be attracted to the positively charged *anode.* Ions are already present in solid *ionic compounds,* but they are in fixed positions and hence cannot move. They are simply "released" from the crystal when the compound dissolves, and thereby become mobile. An example is sodium chloride, NaCl, which may be written $Na^+Cl^-(s)$ to emphasize its character as an ionic solid:

$$Na^+Cl^-(s) \xrightarrow{H_2O} Na^+(aq) + Cl^-(aq) \quad (20.1)$$

With some *molecular compounds* (containing neutral molecules, instead of ions, in their pure state), ions are formed by reaction of the solute with water. For example, hydrogen chloride gas, HCl(g), reacts with water and yields hydrochloric acid:

$$HCl(g) + H_2O(\ell) \rightarrow H_3O^+(aq) + Cl^-(aq) \quad (20.2)$$

In dilute hydrochloric acid, this reaction is virtually complete, and no appreciable number of neutral HCl molecules are present.

Solutions that contain a large number of ions are *good conductors.* The solutions, and the solutes that produce them, are called **strong electrolytes.** Hydrochloric acid is a strong acid. As stated above, its ionization is essentially complete, yielding a large number of ions. Similarly, if you dissolve solid potassium hydroxide, KOH, in water, ionization will be complete and the resultant solution will be a good conductor. There are relatively few strong acids and bases. All *soluble* salts completely ionize when dissolved in water and hence yield strongly conducting solutions.

When acetic acid is dissolved in water, an equilibrium is reached:

$$CH_3COOH(aq) + H_2O \rightleftarrows CH_3COO^-(aq) + H_3O^+(aq) \quad (20.3)$$

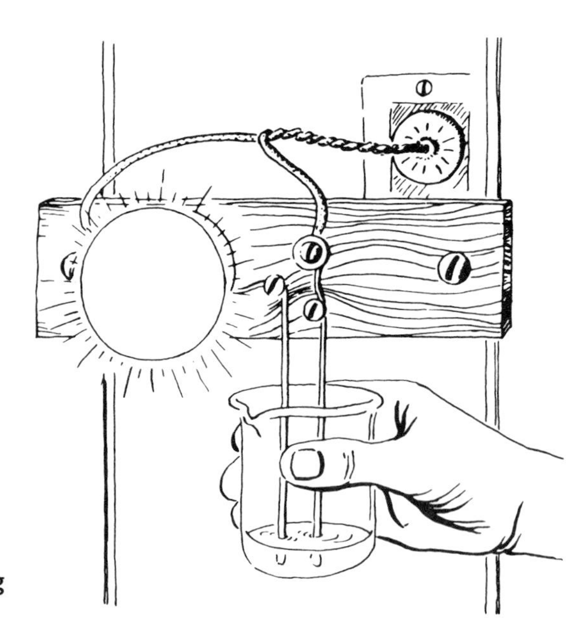

Figure 20–1. Conductivity-sensing apparatus.

Of the original neutral acetic acid molecules, only a small fraction ionizes. Hence, the resulting solution is a *poor conductor*. Most organic acids and bases ionize only to a small extent. Solutes that ionize only slightly and solutions that contain relatively few ions are referred to as **weak electrolytes.**

A third class of solutes includes molecular compounds that do not ionize when dissolved in water. The most minute solute particle remains a neutral molecule. Consequently, such solutions do not conduct an electric current. Solutes whose solutions are *nonconductors* are referred to as **nonelectrolytes.**

Solution conductivity can be detected by a conductivity-sensing apparatus such as the one shown in Figure 20–1. If the electrodes are immersed in a solution containing mobile ions, the ions conduct "current," and the bulb lights. If the solution contains no ions, no current will flow, and the bulb does not light. The flow or nonflow of current is therefore a clear indication of the presence or absence of ions. The intensity of light produced (which is proportional to the magnitude of current) is a qualitative measure of the number of ions present.

In this experiment, you will observe the conductivity of several solutions. You will be asked to make classifications, based on your experimental observations, of whether the substances tested are strong electrolytes, weak electrolytes, or nonelectrolytes.

SAFETY PRECAUTIONS

Do not, at any time, handle the conductivity-testing apparatus unless specifically directed by your instructor. Never touch both electrodes at the same time; if the apparatus is plugged in, a severe electrical shock may result.

PROCEDURE

The instructor will set up a conductivity-sensing apparatus. This apparatus will be used to test the conductivity of several solutions. Classify each solution in the following list as a good conductor, a poor conductor, or a nonconductor, and record your classification in the space provided on the report sheet. Also, answer the corresponding questions in the Discussion section of the report sheet as the experiment progresses.

1. Deionized water
2. Tap water

Explain the difference between the conductivities of deionized water and tap water.

3. Solid NaCl
4. 1.0 M NaCl (salt of a strong acid and strong base). NaCl is an ionic solid.

Explain why solid NaCl did or did not conduct. Would you predict that molten NaCl (pure NaCl in liquid state) would or would not conduct? Why?

5. Glacial acetic acid, CH_3COOH or $HC_2H_3O_2$ (pure acetic acid, no water present)
6. 1.0 M $HC_2H_3O_2$

How would you explain the difference in conductivity between pure acetic acid and dissolved acetic acid?

7. 1.0 M HCl

Compare the conductivities of Steps 6 and 7. Explain the difference between the two acids.

8. 1.0 M NaOH
9. 1.0 M $NH_3(aq)$ (also referred to as NH_4OH)

How would you explain the difference in the observed conductivities?

10. 1.0 M $NaC_2H_3O_2$ (salt of a weak acid and strong base)
11. 1.0 M $NH_4C_2H_3O_2$ (salt of a weak acid and weak base)
12. 1.0 M NH_4Cl (salt of a strong acid and weak base)

Did you find any substantial difference between the conductivities of these solutions? Did they differ from the conductivities of 1.0 M NaCl? Did you find that it made a difference in conductivity whether the soluble salt was formed from a weak or strong acid and a strong or weak base?

13. Solid dextrose (sugar), $C_6H_{12}O_6$
14. 1.0 M dextrose (sugar)

Based on the conductivity of the sugar solution, would you classify it as a molecular solution or an ionic solution? Does it behave like pure water did, or is it similar to a salt solution? Explain your answer. Would you predict that molten sugar is a good conductor, weak conductor, or nonconductor? Can you identify from this experiment another molten compound that behaves as molten sugar should behave, and for the same reason?

15. $BaSO_4$ in water
16. Equal volumes of 1.0 M $BaCl_2$ and 1.0 M Na_2SO_4

Was there any visible evidence of a chemical reaction? Write the conventional equation as well as the net ionic equation for the reaction. Based on your observations in Steps 3 and 15, explain the conductivity behavior of the mixture.

Experiment 20

Work Page

NAME ....

DATE .... SECTION ....

Number	Substance	Conductor (Good, poor, nonconductor)	Electrolyte		
			Strong	Weak	Non
1.	Deionized water				
2.	Tap water				
3.	Solid NaCl				
4.	1.0 M NaCl				
5.	Glacial $HC_2H_3O_2$				
6.	1.0 M $HC_2H_3O_2$				
7.	1.0 M HCl				
8.	1.0 M NaOH				
9.	1.0 M NH_3				
10.	1.0 M $NaC_2H_3O_2$				
11.	1.0 M $NH_4C_2H_3O_2$				
12.	1.0 M NH_4Cl				
13.	Solid dextrose				
14.	1.0 M dextrose				
15.	$BaSO_4$ in water				
16.	1.0 M $BaCl_2$ and 1.0 M Na_2SO_4				

DISCUSSION

1; 2: Explain the difference between the conductivities of deionized water and tap water.

3; 4: Did solid NaCl conduct? Yes ____________ No ____________ . Explain why or why not.

Would you predict that molten NaCl would ____________ or would not ____________ conduct? Explain your reasoning.

5; 6: Explain the reason for the difference in conductivities.

6; 7: The acid having higher conductivity is ________________________ . Give an explanation for the difference.

8; 9: Compare the two bases and explain the difference in conductivities.

Experiment 20
Work Page

NAME ....

DATE .... SECTION ....

10–12: Were the conductivities different? Yes ____________ No ____________ . Give reasons for your observations.

13; 14: Sugar dissolved in water gives an ionic ____________ a molecular ____________ solution, which resembles water ____________ a salt solution ____________ . Molten sugar would be a good ____________ weak ____________ non ____________ conductor. Molten sugar is expected to conduct in the same way as these other liquid compounds:

15; 16: Evidence of a chemical reaction: ____.

Conventional equation: ____.

Net ionic equation: ____.

Explanation for conductivity behavior:

Experiment 20

Report Sheet

NAME ....

DATE .... SECTION ....

Number	Substance	*Conductor (Good, poor, nonconductor)*	*Electrolyte*		
			Strong	*Weak*	*Non*
1.	Deionized water				
2.	Tap water				
3.	Solid NaCl				
4.	1.0 M NaCl				
5.	Glacial $HC_2H_3O_2$				
6.	1.0 M $HC_2H_3O_2$				
7.	1.0 M HCl				
8.	1.0 M NaOH				
9.	1.0 M NH_3				
10.	1.0 M $NaC_2H_3O_2$				
11.	1.0 M $NH_4C_2H_3O_2$				
12.	1.0 M NH_4Cl				
13.	Solid dextrose				
14.	1.0 M dextrose				
15.	$BaSO_4$ in water				
16.	1.0 M $BaCl_2$ and 1.0 M Na_2SO_4				

DISCUSSION

1; 2: Explain the difference between the conductivities of deionized water and tap water.

3; 4: Did solid NaCl conduct? Yes ____________ No ____________ . Explain why or why not.

Would you predict that molten NaCl would ____________ or would not ____________ conduct? Explain your reasoning.

5; 6: Explain the reason for the difference in conductivities.

6; 7: The acid having higher conductivity is ________________________ . Give an explanation for the difference.

8; 9: Compare the two bases and explain the difference in conductivities.

Experiment 20

Report Sheet

NAME . . . .

DATE . . . . SECTION . . . .

10–12: Were the conductivities different? Yes _____________ No _____________ . Give reasons for your observations.

13; 14: Sugar dissolved in water gives an ionic _____________ a molecular _____________ solution, which resembles water _____________ a salt solution _____________ . Molten sugar would be a good _____________ weak _____________ non _____________ conductor. Molten sugar is expected to conduct in the same way as these other liquid compounds:

15; 16: Evidence of a chemical reaction: ____.

Conventional equation: ____.

Net ionic equation: ____.

Explanation for conductivity behavior:

Experiment 20

Advance Study Assignment

NAME ....

DATE .... SECTION ....

1) What is the difference between a strong electrolyte and a weak electrolyte?

2) What do we mean by ionic compound and molecular compound?

3) How does a conductivity-sensing device, such as the one used in this experiment, indicate whether a solution contains ionic or molecular species?

Experiment 21

Net Ionic Equations: A Study Assignment

Performance Goals

21–1 Distinguish between a conventional equation, an ionic equation, and a net ionic equation.

21–2 Identify spectators in an ionic equation.

21–3 Given information from which you may write the conventional equation for one of the following types of reactions in aqueous solution, write the net ionic equation:

- (a) Ion combination reactions that produce a precipitate;
- (b) Ion combination reactions that produce a molecular product;
- (c) Reactions that yield a gas;
- (d) Oxidation–reduction reactions that may be described by a "single replacement" equation.

CHEMICAL OVERVIEW

The conventional equation by which a chemical change is described is adequate for most purposes, but for reactions of ionic compounds in aqueous (water) solution, the typical molecular equation has serious shortcomings. Specifically, a conventional equation may indicate formulas of reactants and products that are not present, and omit completely the formulas of the ions that are the real reactants and products. If the substances in the conventional equation that are actually present as dissociated ions are written in the form of their ions, the result is an **ionic equation.** Frequently the same ions appear on both sides of an ionic equation. Though they are present at the scene of a chemical reaction, these ions experience no chemical change themselves. They are called **spectator ions,** or simply **spectators.** Eliminating the spectators from an ionic equation leaves the **net ionic equation,** an equation that includes only the substances that actually participate in the reaction.

To illustrate these equations, consider the reaction between solutions of lead(II) nitrate and sodium iodide. When the two clear solutions are combined, insoluble lead(II) iodide precipitates. The conventional equation is

$$Pb(NO_3)_2(aq) + 2\ NaI(aq) \rightarrow PbI_2(s) + 2\ NaNO_3(aq) \qquad (21.1)$$

Three of the compounds in the equation are soluble ionic salts that dissociate into ions when dissolved. In other words, no real solute particles having the formulas $Pb(NO_3)_2$, NaI, or $NaNO_3$ are actually present in the reaction vessel. The species actually present are the ions resulting from those species. These ions, both their identity and number, can be found by writing the equation for the dissociation of that number of solute species appearing in the conventional equation:

$$Pb(NO_3)_2(aq) \rightarrow Pb^{2+}(aq) + 2\ NO_3^-(aq) \qquad (21.2)$$

$$2\ NaI(aq) \rightarrow 2\ Na^+(aq) + 2\ I^-(aq) \qquad (21.3)$$

$$2\ NaNO_3(aq) \rightarrow 2\ Na^+(aq) + 2\ NO_3^-(aq) \qquad (21.4)$$

Replacing the compound formulas of dissolved substances in the conventional equation with the ions really present gives the ionic equation:

$$Pb^{2+}(aq) + 2\ NO_3^-(aq) + 2\ Na+(aq) + 2\ I^-(aq) \rightarrow PbI_2(s) + 2\ Na^+(aq) + 2\ NO_3^-(aq) \quad (21.5)$$

The ionic equation contains two sodium ions and two nitrate ions on both sides. These are the spectators that do not undergo any change, even though they are in the solution where chemical change is occurring. Eliminating the spectators from the ionic equation gives the net ionic equation:

$$Pb^{2+}(aq) + 2\ I^-(aq) \rightarrow PbI_2(s) \quad (21.6)$$

The net ionic equation isolates the two reactants that actually change chemically and the single new substance produced.

The above example does more than simply illustrate a net ionic equation. It also shows how that equation is developed. There are three steps in the procedure:

1. Write the conventional equation, including designations of state or solutions, (s) for solid, (ℓ) for liquid, (g) for gas, and (aq) for dissolved species (Equation 21.1). Be sure the equation is balanced.*
2. Write the ionic equation, replacing those species that ionize in aqueous solution with ions actually present (Equation 21.5). Check to be sure the equation is still balanced, both in atoms and in electrical charge.
3. Write the net ionic equation by eliminating the spectators (Equation 21.6). If necessary, reduce coefficients to their lowest integral (whole number) values. Check to be sure the equation is still balanced, both in atoms and electrical charge.

Regardless of what you may or may not have been doing up to this point in your chemistry course, it is particularly important that you include designations of state or solution in your conventional equation. These designations identify the formulas that *may* be rewritten in the form of dissociated ions in the ionic equation. *Under no circumstances do you change the formula of a solid, liquid, or gas; those substances appear in the ionic and net ionic equations in exactly the same form as they appear in the conventional equation.* It follows that you must have sufficient information in the description of the reaction to determine the proper designations for all substances. Reactant designations are determined by careful reading of the description of the reaction. Product designations may or may not be indicated. It is generally left to you to decide whether a product formed by the combination of ions is or is not soluble, and, even if it is soluble, whether it does or does not dissociate into ions.

Solubility. The solubility of an ion combination product may be determined by referring to Table 21–1. Directions for its use appear beneath the table. Notice that the intersection of the horizontal lead(II) ion line with the vertical iodide ion column contains the letter "s," indicating that lead(II) iodide is an insoluble solid, written as $PbI_2(s)$ in an equation. The intersection of the horizontal sodium ion line with the vertical nitrate ion column shows "aq," indicating that sodium nitrate is soluble in water and is written $NaNO_3(aq)$ in an equation. It is worth noting that all entries in the sodium ion line and the nitrate ion column are designated "aq," showing that all sodium salts and all nitrate salts included in the table are soluble. These and several other useful generalizations are listed in the following "solubility rules":

1. All ammonium and alkali metal salts are soluble.
2. All nitrates are soluble.
3. All chlorides and bromides are soluble except those of Ag^+, Hg_2^{2+}, and Pb^{2+} ions.
4. All sulfates are soluble except those of Ca^{2+}, Sr^{2+}, Ba^{2+}, Hg_2^{2+}, Hg^{2+}, Pb^{2+}, and Ag^+ ions.
5. All carbonates are insoluble except ammonium carbonate and the carbonates of the alkali metals.

* If the purpose is simply to write the net ionic equation for a reaction, it is not essential that the conventional and ionic equations be balanced. Most instructors, however, require balancing all three equations. That procedure will be followed in this book; you should follow the procedure recommended by your instructor.

Table 21-1 Solubilities of Ionic Compounds*

Anions / *Cations*	*Acetate*	*Bromide*	*Carbonate*	*Chlorate*	*Chloride*	*Fluoride*	*Hydrogen Carbonate*	*Hydroxide*	*Iodide*	*Nitrate*	*Nitrite*	*Phosphate*	*Sulfate*	*Sulfide*	*Sulfite*
Aluminum	s	aq		aq	aq	s		s	—	aq		s	aq	—	
Ammonium	aq	aq	aq	aq	aq	aq	aq	—	aq	aq	aq	aq	aq	aq	aq
Barium	—	aq	s	aq	aq	s		aq	aq	aq	aq	s	s	—	s
Calcium	aq	aq	s	aq	aq	s		s	aq	aq	aq	s	s	s	s
Cobalt(II)	aq	aq	s	aq	aq	—		s	aq	aq		s	aq	s	s
Copper(II)	aq	—			aq	aq		s		aq		s	aq	s	
Iron(II)	aq	aq	s		aq	s		s	aq	aq		s	aq	s	s
Iron(III)	—	aq			aq	s		s	—	aq		s	aq	—	
Lead(II)	aq	s	s	aq	s	s		s	s	aq	aq	s	s	s	s
Lithium	aq	aq	aq	aq	aq	aq	aq	aq	aq	aq	aq	s	aq	aq	
Magnesium	aq	aq	s	aq	aq	s		s	aq	aq	aq	s	aq	—	aq
Nickel		aq	s	aq	aq	aq		s	aq	aq		s	aq	s	s
Potassium	aq	aq	aq	aq	aq	aq	aq	aq	aq	aq	aq	aq	aq	aq	aq
Silver	s	s	s	aq	s	aq		—	s	aq	s	s	s	s	s
Sodium	aq	aq	aq	aq	aq	aq	aq	aq	aq	aq	aq	aq	aq	aq	aq
Zinc	aq	aq	s	aq	aq	aq		s	aq	aq		s	aq	s	s

* To determine the solubility of an ionic compound, locate the intersection of the horizontal row for the cation and the vertical column for the anion. An "aq" in that box indicates that the compound is soluble in water to a molarity of 0.1 or more at 20°C. An "s" indicates that the compound is not soluble to that concentration, but remains in the solid state or precipitates if the ions are combined. A blank space indicates lack of data, and a dash (—) identifies an unstable substance.

6. All hydroxides are insoluble except those of the alkali metals, $Ba(OH)_2$ and $Sr(OH)_2$.
7. All sulfides are insoluble except those of the alkali metals and ammonium sulfide.

Molecular Products — Weak Acids. Sometimes the hydrogen ion from a strong acid will combine with the anion of a second reactant to form a molecular product. If the anion is the hydroxide ion, the molecular product is water, $H_2O(\ell)$, which appears in that form in all conventional and ionic equations. Another molecular product forms when the hydrogen ion from a strong acid combines with the anion of a weak acid. These acids ionize only slightly, and when writing ionic equations are considered to exist in molecular form. Hydrofluoric acid, HF, for example, happens to be an extremely corrosive acid, but it is also a weak acid because the individual solute particles in its water solution are predominantly unionized HF molecules. Its proper designation is therefore HF(aq) in ionic equations as well as conventional equations. In other words, *a weak acid is not separated into ions in writing an ionic equation, but is written in molecular form.*

All of this makes it necessary for you to distinguish between weak acids and strong acids. Fortunately, there are only a few strong acids — few enough for you to memorize them easily. They are the three most common acids, sulfuric (H_2SO_4), nitric (HNO_3), and hydrochloric (HCl), and three other acids containing halogens, hydrobromic (HBr), hydroiodic (HI), and perchloric ($HClO_4$). If any of these six acids appears in a conventional equation, it must be broken down to ions in an ionic equation. All other acids are written in molecular form in both conventional and ionic equations.*

Unstable Ion Combination Products. There are three common substances that are unstable in water solution, and must be written in both the conventional and ionic equations in the form of their final decomposition products. These are carbonic acid, H_2CO_3, sulfurous acid, H_2SO_3, and what is commonly called "ammonium hydroxide," whose formula is usually written NH_4OH. Carbonic and sulfurous acids break down into water plus carbon dioxide or sulfur dioxide, respectively. Carbon dioxide is relatively insoluble and bubbles out of the solution as a gas, $CO_2(g)$. Sulfur dioxide is relatively soluble, and is written as $SO_2(aq)$ in the equation. "Ammonium hydroxide," NH_4OH, never really forms from NH_4^+ and OH^- ions; instead, an equilibrium is reached, $NH_4^+(aq) + OH^-(aq) \rightleftarrows NH_3(aq) + H_2O(\ell)$, in which the products are by far the predominant species and are written in molecular form in all equations.

EXAMPLES

The following examples are in the form of a program in which you learn by answering a series of questions. Obtain an opaque shield — a piece of cardboard, or a folded piece of paper you cannot see through — that is wide enough to cover this page. In each example, place the shield on the book page so it covers everything beneath the first dotted line that runs across the page. Read to that point, and write in the space provided whatever is asked. Then lower the shield to the next dotted line. The material exposed will begin with the correct response to the question you have just answered. Compare this answer to yours, looking back to correct any misunderstanding if the two are different. When you fully understand the first step, read to the next dotted line and proceed as before.

Example 1

Write the net ionic equation for the reaction between solutions of barium bromide and ammonium sulfate.

Begin with the conventional equation. You will have to predict the products of the reaction, which is not difficult if you recall that the reaction between two ionic substances is frequently described by a "double displacement" type of equation, in which the positive ion of one reactant combines with the negative ion of the other, and vice versa — a "change partners" type of equation. Be sure to include state or solution designations, using Table 21–1 to determine the proper designations for the products.

. . . .

1a. $BaBr_2(aq) + (NH_4)_2SO_4(aq) \rightarrow BaSO_4(s) + 2\ NH_4Br(aq)$

The (aq) designations for the reactants are indicated by the word "solutions" in the statement of the question. The intersection of the barium ion line and sulfate column line in Table 21–1 has an "s" in it, indicating that barium sulfate is a solid (precipitate), and therefore its designation in the equation is (s). The ammonium ion and bromide ion intersection shows "aq," indicating that the compound is soluble in water.

Now write the ionic equation by replacing those species that are in solution, designated (aq), by the ionic particles actually present in the solution. Remember, only those substances designated (aq) may be changed in this step. Be sure your equation is balanced.

* Some fairly common acids, notably phosphoric, oxalic, and chloric acids, are "borderline" acids that might be considered either strong or weak, depending on some arbitrary classification standard. We will avoid these acids in this exercise.

. . . .

1b. $Ba^{2+}(aq) + 2\ Br^-(aq) + 2\ NH_4^+(aq) + SO_4^{2-}(aq) \rightarrow BaSO_4(s) + 2\ Br^-(aq) + 2\ NH_4^+(aq)$

The $BaSO_4(s)$, a solid, remains in the same form in the ionic equation as it appeared in the conventional equation. The other substances, all ionic compounds, dissociate into their ions.

From here all you must do is to cross out the spectators and you have the net ionic equation.

. . . .

1c. $Ba^{2+}(aq) + SO_4^{2-}(aq) \rightarrow BaSO_4(s)$

A final check on the equation shows one barium ion on each side, one sulfate ion on each side, and a total charge of zero on both sides. The importance of the charge balance will appear shortly.

The next example is similar, but slightly different. . . .

Example 2

Write all equations that result in the net ionic equation for the reaction between solutions of silver nitrate and magnesium chloride.

You know the procedure for all three steps. Proceed, but be careful on the last step. . . .

. . . .

2a. $2\ AgNO_3(aq) + MgCl_2(aq) \rightarrow 2\ AgCl(s) + Mg(NO_3)_2(aq)$

$2\ Ag^+(aq) + 2\ NO_3^-(aq) + Mg^{2+}(aq) + 2\ Cl^-(aq) \rightarrow 2\ AgCl(s) + Mg^{2+}(aq) + 2\ NO_3^-(aq)$

$2\ Ag^+(aq) + 2\ Cl^-(aq) \rightarrow 2\ AgCl(s)$, or simply $Ag^+(aq) + Cl^-(aq) \rightarrow AgCl(s)$

The net ionic equation that results from the elimination of the magnesium and nitrate ion spectators has coefficients of 2 for all species. Step 3 in the procedure for writing net ionic equations (page 246) states that all coefficients should be expressed in their lowest terms. Dividing the first equation in the last line above by 2 yields the second equation.

At some point you should become aware of the possible combination of solutions that leads to the conditions in the next example.

Example 3

Write the net ionic equation for the reaction, if any, that occurs when a solution of sodium nitrate is poured into a solution of ammonium sulfate.

Proceed, but when you come to something you haven't seen before, think about it. . . .

. . . .

3a. $2\ NaNO_3(aq) + (NH_4)_2SO_4(aq) \rightarrow Na_2SO_4(aq) + 2\ NH_4NO_3(aq)$

$2\ Na^+(aq) + 2\ NO_3^-(aq) + 2\ NH_4^+(aq) + SO_4^{2-}(aq) \rightarrow$
$2\ Na^+(aq) + SO_4^{2-}(aq) + 2\ NH_4^+(aq) + 2\ NO_3^-(aq)$

There is no net ionic equation because every species in the ionic equation is a spectator. In other words, there is no reaction, no chemical change, no reactant destroyed, no new substance formed. The final mixture contains the four ions that were initially divided between two solutions. When you encounter a question such as this, simply state, "No reaction."

A neutralization reaction is one between an acid and a base, yielding water and a salt as its products. The next example gives the net ionic equation that describes all neutralizations between solutions of a strong acid and a strong base.

Example 4

Write the net ionic equation for the reaction between aqueous potassium hydroxide and sulfuric acid.

The procedure is the same as before, but there will be a small difference that shows up in your state designations. Complete the equation.

. . . .

4a. $H_2SO_4(aq) + 2\ KOH(aq) \rightarrow 2\ H_2O(\ell) + K_2SO_4(aq)$

$2\ H^+(aq) + SO_4^{2-}(aq) + 2\ K^+(aq) + 2\ OH^-(aq) \rightarrow 2\ H_2O(\ell) + SO_4^{2-}(aq) + 2\ K^+(aq)$

$2\ H^+(aq) + 2\ OH^-(aq) \rightarrow 2\ H_2O(\ell)$, or simply $H^+(aq) + OH^-(aq) \rightarrow H_2O(\ell)$

The difference in state designation is that water is a liquid, designated (ℓ), rather than a solid precipitate as in Examples 1 and 2.

There is a common error that frequently appears in writing the net ionic equation for a reaction involving sulfuric acid, as in Example 4. In writing the ionic equation the acid is incorrectly broken into the sulfate ion and an $H_2^+(aq)$ ion, which does not exist. The formula for an aqueous hydrogen ion is $H^+(aq)$, whether it comes from H_2SO_4, HCl, or any other acid. The fact that elemental hydrogen is diatomic, H_2, has nothing to do with the aqueous hydrogen ion.

Example 5

Write the net ionic equation for the reaction between hydrochloric acid and solid barium hydroxide.

The reaction is very similar to Example 4, but you must read carefully. . . .

. . . .

5a. $2\ HCl(aq) + Ba(OH)_2(s) \rightarrow 2\ H_2O(\ell) + BaCl_2(aq)$

$2\ H^+(aq) + 2\ Cl^-(aq) + Ba(OH)_2(s) \rightarrow 2\ H_2O(\ell) + Ba^{2+}(aq) + 2\ Cl^-(aq)$

$2\ H^+(aq) + Ba(OH)_2(s) \rightarrow 2\ H_2O(\ell) + Ba^{2+}(aq)$

In this example it is stated that *solid* barium hydroxide is the reactant, even though barium hydroxide is sufficiently soluble to be classified (aq) in Table 21–1. Watch for words in the statement of a reaction that indicate the state of a reactant, and be sure to use that state in the equation.

The next neutralization reaction yields a slightly different net ionic equation.

Example 6

Write the net ionic equation for the reaction between hydrochloric acid and a solution of sodium fluoride.

Proceed as before, but remember the discussion of weak acids on page 247.

. . . .

6a. $HCl(aq) + NaF(aq) \rightarrow HF(aq) + NaCl(aq)$

$H^+(aq) + Cl^-(aq) + Na^+(aq) + F^-(aq) \rightarrow HF(aq) + Na^+(aq) + Cl^-(aq)$

$H^+(aq) + F^-(aq) \rightarrow HF(aq)$

In this case the product HF(aq) is a weak acid — recognized as such because it is not one of the six strong acids — and therefore does not separate into ions.

The base used to neutralize an acid is not necessarily a hydroxide, as the next example shows.

Example 7

Write the net ionic equation for the reaction between hydrochloric acid and a solution of sodium hydrogen carbonate.

Again proceed, but this time remember the discussion of unstable ion combination products on page 248.

. . . .

7a. $HCl(aq) + NaHCO_3(aq) \rightarrow H_2O(\ell) + CO_2(g) + NaCl(aq)$

$H^+(aq) + Cl^-(aq) + Na^+(aq) + HCO_3^-(aq) \rightarrow H_2O(\ell) + CO_2(g) + Na^+(aq) + Cl^-(aq)$

$H^+(aq) + HCO_3^-(aq) \rightarrow H_2O(\ell) + CO_2(g)$

This time the ion combination product is unstable carbonic acid, H_2CO_3, which decomposes into water and carbon dioxide.

The reaction between an active metal and an acid releases hydrogen gas and leaves a solution of a salt behind. This kind of oxidation–reduction reaction is described by a conventional equation that is sometimes called a single replacement equation, in which one element appears to replace another.

Example 8

Write the net ionic equation for the reaction between zinc and sulfuric acid.

. . . .

8a. $Zn(s) + H_2SO_4(aq) \rightarrow H_2(g) + ZnSO_4(aq)$

$Zn(s) + 2\ H^+(aq) + SO_4^{2-}(aq) \rightarrow H_2(g) + Zn^{2+}(aq) + SO_4^{2-}(aq)$

$Zn(s) + 2\ H^+(aq) \rightarrow H_2(g) + Zn^{2+}(aq)$

In the conventional equation it looks as if zinc replaces hydrogen to form a compound with the sulfate ion. The net ionic equation shows the true character of the reaction, in which two electrons move from a zinc atom to two hydrogen ions. It is this transfer of electrons that classifies the reaction as an oxidation–reduction reaction.

Example 8 differs from all of the earlier examples in that each side of the equation has a net electrical charge, instead of having all charges cancel out to zero. The requirement is that the charges be *balanced,* but not necessarily at zero. If you balance the conventional equation and correctly represent the number of ions in the ionic equation that come from the conventional equation, the charges in the net ionic equation will already be balanced. They should be checked, however, and not assumed. If the charge is not balanced, something is wrong someplace in the three steps. It is possible to write a net ionic equation that is balanced in atoms but not in charge. The final example is such a reaction.

Example 9

Write the net ionic equation for the reaction between copper and a solution of silver nitrate.

The conventional equation for this oxidation–reduction reaction between copper and aqueous silver nitrate is a single replacement equation in which the copper(II) ion is produced. This hint should enable you to write the three equations.

. . . .

9a. $Cu(s) + 2\ AgNO_3(aq) \rightarrow 2\ Ag(s) + Cu(NO_3)_2(aq)$

$Cu(s) + 2\ Ag^+(aq) + 2\ NO_3^-(aq) \rightarrow 2\ Ag(s) + Cu^{2+}(aq) + 2\ NO_3^-(aq)$

$Cu(s) + 2\ Ag^+(aq) \rightarrow 2\ Ag(s) + Cu^{2+}(aq)$

An error in the first or second step could have produced $Cu(s) + Ag^+(aq) \rightarrow Ag(s) + Cu^{2+}(aq)$ as a net ionic equation that is balanced in atoms. The net charge on the left side of the equation is 1+, however, and the charge on the right side is 2+. The equation is therefore not balanced, even though both copper and silver check out.

DEMONSTRATION (OPTIONAL)

Observe the following process carried out by the instructor, and write the net ionic equation for the reaction. Also, explain the reason for the changes during the titration.

Procedure. Pour 30 to 35 mL of clear, saturated barium hydroxide solution (carefully decanted from a bottle containing excess solute on the bottom) into a 400-mL beaker. Add about 100 mL of deionized water and 3 drops of phenolphthalein indicator. Immerse the electrodes of a conductivity-sensing device (see Figure 20–1) into the solution.

In a solution, the current carrying species are ions; hence, good conduction indicates a great number of ions. Conversely, poor conduction shows that a relatively small number of ions are present. Observe whether the barium hydroxide solution in the beaker is a good or a poor conductor of electricity.

From a buret, slowly add a 1M sulfuric acid solution to the contents of the beaker. Stir the solution continuously and observe any changes that occur.

Experiment 21

Report Sheet

NAME ....

DATE .... SECTION ....

Instructions: For each reaction described below, write the conventional equation, ionic equation, and net ionic equation. Include designations of state or solution in each equation.

1) Lead(II) nitrate and magnesium sulfate solutions are combined.

2) Barium metal is dropped into hydrochloric acid.

3) Potassium hydroxide solution reacts with nitric acid.

4) Zinc chloride solution is poured into a solution of ammonium carbonate.

5) Ammonium chloride and potassium hydroxide solutions are combined.

6) Magnesium chloride solution is mixed with nickel nitrate solution.

7) Sulfuric acid reacts with a solution of magnesium acetate, $Mg(C_2H_3O_2)_2$.

8) Zinc reacts with a solution of nickel sulfate.

9) Cobalt(II) sulfate and lithium sulfide solutions are combined.

10) Potassium nitrite solution is added to sulfuric acid.

11) Solid nickel carbonate is dropped into nitric acid.

12) Iron(III) chloride and aluminum sulfate solutions are mixed.

13) Hydrochloric acid is poured into a solution of lithium carbonate.

Experiment 21

Report Sheet

NAME ....

DATE .... SECTION ....

14) Hydrogen is released when sodium reacts with water.

15) Hydroiodic acid reacts with a solution of ammonium sulfite.

16) Solid copper(II) hydroxide is "dissolved" by hydrochloric acid.

17) Sodium hydroxide solution is poured into a solution of cobalt(II) chloride.

18) Calcium metal reacts with a solution of iron(II) bromide.

19) Hydrochloric acid reacts with a solution of sodium butyrate, $NaC_4H_7O_2$.

20) Aluminum bromide and ammonium fluoride solutions are combined.

DEMONSTRATION QUESTIONS (OPTIONAL)

1) What were the color and conductivity of the initial barium hydroxide and phenolphthalein solution?

2) What changes occurred during the titration?

3) Why did the pink color disappear?

4) Why did the light go out?

5) Why did the light come back on?

6) Write the conventional and the net ionic equations for the reaction.

Experiment 21

NAME ....

Advance Study Assignment

DATE .... SECTION ....

1) An ionic equation has $2H^{+}(aq)$ on the left side and $H_2(g)$ on the right. Are these species spectators? Explain.

2) Which of the following substances are *not* ionized in an aqueous solution? Explain.

a) HI; b) KBr; c) H_2SO_4; d) HNO_2; e) AgCl; f) $Ca(NO_3)_2$

3) Consider the statement: after balancing a net ionic equation, the sum of the charges on each side of the equation should equal zero. Is this true or false? Explain.

Experiment 22®

The Chemistry of Some Household Products

Performance Goals

22–1 Perform tests to confirm the presence of known ions in certain solids.
22–2 Perform tests to confirm the presence of these ions in household products.
22–3 Analyze and identify an unknown solid.

CHEMICAL OVERVIEW

In everyday life we encounter a large number of chemicals. In fact, everything that surrounds us is a chemical. The air we breathe is a mixture of gaseous elements, oxygen and nitrogen. The food we eat contains carbohydrates, proteins, large molecular mass vitamins, etc. You may not think about it, but you handle all kinds of chemicals every day.

Qualitative analysis is used to identify components of a solution or solid (see Experiment 14). A reagent that causes an easily recognized reaction with a particular ion present is added to a sample of the unknown. If the reaction occurs, the ion is present; if the reaction does not occur, the ion is absent.

In this experiment you will perform tests on known compounds that show the presence of certain ions. Once you have become familiar with these specific reactions, you will perform the same tests on some common household products. Finally, you will be given an unknown compound. By performing a series of tests, you will determine which of the following ions is present in your unknown: Cl^-, NH_4^+, SO_4^{2-}, HCO_3^-, PO_4^{3-}.

SAFETY PRECAUTIONS AND DISPOSAL METHODS

Acids and bases are corrosive and contact with your skin should be avoided. Any spilled acid or base should be washed off promptly. Be sure to wear goggles or safety glasses while performing this experiment.

Dispose of excess solids as directed by your instructor. Solutions can be poured into the sink and rinsed down with cold water.

PROCEDURE

1. TABLE SALT, NaCl

The presence of chloride ions, Cl^-, can be detected by reacting the dissolved substance with silver nitrate, $AgNO_3$. A white precipitate of silver chloride, AgCl, will form. This reaction is typical of all solutions that contain the Cl^- ion.

Place a small amount — about 3 mm on the tip of a spatula — of table salt in a test tube. Dissolve the solid in about 10 drops of deionized water. Add 2 drops of 1 M HNO_3. Then add 2 or 3 drops of 0.1 M $AgNO_3$. Record your observations.

2. GARDEN FERTILIZERS

Some of the active ingredients of ordinary garden fertilizers are ammonium salts. These compounds are the source of nitrogen, an element essential for the growth of plants. When a strong base, such as sodium hydroxide, NaOH, is added to a compound containing the NH_4^+ ion, gaseous ammonia, NH_3, is liberated. This can be detected by a piece of red litmus paper that has been moistened with deionized water. The paper will turn blue if NH_3 is present.

A) Pour about 10 drops of 1 M NH_4Cl into a small test tube. Add about 10 drops of 3 M NaOH. Hold a piece of moist red litmus paper in the mouth of the test tube. Do not allow the paper to come into contact with the side of the test tube, since it may have NaOH on it. Record your observations. If you notice no change, gently warm the test tube in a hot-water bath and check with litmus paper again.

B) Place a small amount of garden fertilizer into a test tube. Add about 10 drops of 3 M NaOH to the solid. Hold a moist strip of red litmus paper in the mouth of the test tube. Record your observations.

3. EPSOM SALT, $MgSO_4 \cdot 7H_2O$

Epsom salt can be purchased in any drugstore. It is commonly used to prepare soothing baths, and it is sometimes used as a purgative. It has quite a bitter taste. If a solution of barium chloride, $BaCl_2$, is added to a solution of Epsom salt, a finely divided white precipitate of barium sulfate, $BaSO_4$, will form.

A) Pour about 10 drops of 1 M Na_2SO_4 into a test tube. Add 3 drops of 1 M HCl and 2 or 3 drops of 1 M $BaCl_2$. Record your observations.

B) Place a small amount of Epsom salt into a test tube. Dissolve the solid in about 10 drops of deionized water. Add 3 drops of 1 M HCl and 2 or 3 drops of 1 M $BaCl_2$. Record your observations.

4. BAKING SODA, $NaHCO_3$

Baking soda and baking powder both contain sodium hydrogen carbonate, $NaHCO_3$. When this compound reacts with an acid, gaseous carbon dioxide is produced. This makes bread or cakes "rise." Carbon dioxide is produced whenever compounds containing HCO_3^- or CO_3^{2-} ions react with acids. The formation of bubbles indicates that at least one of these ions is present in a sample.

A) Dissolve a small amount of baking soda in about 10 drops of deionized water. Add about 6 – 8 drops of 1 M HCl. Record your observations. Repeat the test by dissolving a small amount of baking soda directly in HCl. Again record your observations, noting any difference between the two reactions.

B) Take another small amount of solid baking soda and add 10 drops of commercial vinegar to it. Record your observations, noting any difference between this reaction and the ones in Step A.

5. DETERGENTS

One of the common ingredients of laundry detergents and wall-washing compounds is sodium phosphate, Na_3PO_4. The PO_4^{3-} ion can be detected by adding ammonium molybdate $(NH_4)_2MoO_4$ to a dissolved sample. A yellow, powdery precipitate will form. Sometimes gentle heating in a water bath is necessary to hasten the reaction.

A) Pour about 10 drops of 1 M Na_3PO_4 into a test tube and add 3 M HNO_3 until the solution is acidic. (Test by dipping a stirring rod into the solution and touching the wet rod to a strip of blue litmus paper. The solution is acidic if the color changes to red.) Then add 6 – 8 drops of 0.5 M $(NH_4)_2MoO_4$ and place the test tube in a hot-water bath for a few minutes. Record your observations.

B) Repeat the procedure with a small amount of laundry detergent. Dissolve the solid in about 10 drops of deionized water, acidify it, and add the molybdate reagent. Record your observations. Heat the solution in a water bath and note any change, if there is one.

6. IDENTIFICATION OF AN UNKNOWN

Obtain a solid unknown and record its number on your report sheet. Place small portions of the unknown in five separate test tubes. Keep enough unknown to make three additional tests, in case you wish to repeat one or more of the procedures. Dissolve each solid portion in about 10 drops of deionized water and perform the five tests on the separate portions. Your unknown will contain only one compound. Identify on your report sheet the ion you believe to be present.

Experiment 22
Work Page

NAME ....

DATE .... SECTION ....

Substance Tested	
Table salt	
NH_4Cl	
Fertilizer	
Na_2SO_4	
Epsom salt	
Baking soda solution + HCl	
Baking soda solid + HCl	
Baking soda solid + vinegar	
Na_3PO_4	
Detergent	

Unknown Number ________________ Ion present ______________________________________

Describe the experimental observation on which your result is based:

Experiment 22
Report Sheet

NAME ....

DATE .... SECTION ....

Substance Tested	
Table salt	
NH_4Cl	
Fertilizer	
Na_2SO_4	
Epsom salt	
Baking soda solution + HCl	
Baking soda solid + HCl	
Baking soda solid + vinegar	
Na_3PO_4	
Detergent	

Unknown Number ________________ Ion present ________________________________

Describe the experimental observation on which your result is based:

Experiment 22
Advance Study Assignment

NAME ....

DATE .... SECTION ....

1) What would happen if you used tap water instead of deionized water in this experiment?

2) What test would you perform to decide if your fertilizer contains ammonium salts?

3) If you have a white powder that could be either baking soda or table salt, how could you decide which one it is?

4) Why is it dangerous to mix household chemicals indiscriminately?

Experiment 23, Experiment 24

Titration of Acids and Bases: An Introduction

Experiments 23 and 24 are really two parts of a single experiment having a single ultimate objective, namely, the determination of the concentration of an acid solution. This special overview has been written to point out the division of the procedure into two parts, and to furnish you a basis for understanding how these parts are related. In brief, Experiment 24 uses a sodium hydroxide solution of *known* concentration to find the unknown concentration of the acid. The sodium hydroxide solution used in Experiment 24 is prepared, and its concentration is determined, in Experiment 23. It is therefore essential that the solution you prepare in Experiment 23 be kept for use in Experiment 24.

MOLARITY: A CONCENTRATION UNIT

In these experiments you will use **molarity** as your concentration unit. By definition, molarity is the number of moles of solute per liter of solution. So you may see clearly how molarity may be found from experimental data — so you may see what information you require to calculate molarity, and how you use that information — the idea is here presented side by side with a more familiar parallel:

Rate of travel is expressed by speed, the units of which are miles per hour, or miles/hour.

If you were to hike 12.0 miles in 4.00 hours, you would find your average speed by dividing 12.0 miles by 4.00 hours:

$$\frac{12.0 \text{ miles}}{4.00 \text{ hours}} = 3.00 \text{ miles/hour}$$

Your calculation method matches the units in which speed is expressed: if you divide the number of miles travelled in a given trip by the hours taken for the same trip, the result is average speed.

The two essential items you require for calculation of speed are *miles travelled* and *hours in trip.*

Solution concentration is expressed by molarity, the units of which are moles per liter, or moles/liter.

If you were to dissolve 12.0 moles of solute in 4.00 liters of solution, you would find your molarity by dividing 12.0 moles by 4.00 liters:

$$\frac{12.0 \text{ moles}}{4.00 \text{ liters}} = 3.00 \text{ moles/liter}$$

Your calculation method matches the units in which molarity is expressed: if you divide the number of moles in a given sample of solution by the number of liters in the same sample, the result is the molarity of the solution.

The two essential items you require for calculation of molarity are *moles of solute* and *liters of solution.*

The last sentence in the right column sets your objective for both experiments: to find the number of moles of solute in a sample of solution, and the volume of that sample. Dividing one by the other yields the required molarity.

TITRATION: A LABORATORY PROCESS

Titration is the controlled addition of a solution into a reaction vessel from a buret. By means of titration, the volume of solution used may be determined quite precisely. The titration process is used in many analytical determinations, including those involving acid–base reactions.

An **indicator** is a substance used to signal when the titration arrives at the point at which the reactants are stoichiometrically (or chemically) equal, as defined by the reaction equation. For example, in an acid–base titration between sodium hydroxide and hydrochloric acid,

$$NaOH(aq) + HCl(aq) \rightarrow H_2O(\ell) + NaCl(aq),$$

the indicator should tell when the numbers of moles of NaOH and HCl are exactly equal, matching the 1:1 ratio in the equation. For the reaction

$$2NaOH(aq) + H_2SO_4(aq) \rightarrow 2\,H_2O(\ell) + Na_2SO_4(aq),$$

the indicator should tell when the number of moles of NaOH is exactly twice the number of moles of H_2SO_4, this time reflecting the 2:1 molar ratio between the reactants. This point of chemical equality is called the **equivalence point** of the titration. Acid-base indicators send their signal by changing color at or very near the equivalence point of the titration.

A **standard solution** is a solution with a precisely determined concentration. Initially the concentration of a standard solution is determined from a weighed quantity of a **primary standard,** a highly purified reference chemical. A standard solution may be prepared in either of two ways:

1. A primary standard is carefully weighed, dissolved, and diluted accurately to a known volume. Its concentration can be calculated from the data.
2. A solution is made to an approximate concentration and then standardized by titrating an accurately weighed quantity of a primary standard.

Once a solution has been standardized in one reaction, it may be used as a standard solution in subsequent reactions. Thus the standard solution prepared in Experiment 23 will be used in the reaction of Experiment 24 to determine the concentration of an unknown acid.

With this background, we now proceed to Experiments 23 and 24 individually.

Experiment 23

Titration of Acids and Bases — I

Performance Goals

23–1 Given the volume of a solution of known molarity, and the volume to which it is diluted with water, calculate the molarity of the diluted solution.

23–2 Given the approximate molarity and volume of an acid or base solution to be used in a titration, calculate the number of grams of a known solid base or acid required for the reaction.

23–3 Given the volume of a base or acid solution that reacts with a weighed quantity of a primary standard acid or base, calculate the molarity of the base or acid solution.

23–4 Perform acid–base titrations reproducibly.

CHEMICAL OVERVIEW

In this experiment you will prepare a standard solution of sodium hydroxide to be used in Experiment 24. Solid sodium hydroxide has the property of absorbing moisture from the air. It is therefore not possible to weigh sodium hydroxide accurately, which makes it unsuitable as a primary standard. Consequently, you will use the second of the two methods for preparing a standard solution listed on page 270. Your primary standard will be oxalic acid dihydrate, $H_2C_2O_4 \cdot 2\,H_2O$. The reaction between the acid and base is

$$2\,NaOH(aq) + H_2C_2O_4(aq) \rightarrow 2\,H_2O(\ell) + Na_2C_2O_4(aq) \qquad (23.1)$$

Sodium hydroxide will be made available in the laboratory in the form of a solution that is approximately one molar (1 M) in concentration. Note that this is an *approximate* concentration, expressed in *one* significant figure. No calculation based on that concentration can be considered reliable. You will be instructed to dilute a specified quantity of that solution to a larger volume with water, and then to calculate the *approximate* concentration of the diluted solution (see Performance Goal 23–1). This is one of two preliminary calculations in this experiment, and it appears as Question 1 in the Advance Study Assignment. If you turn in your Advance Study Assignment at the beginning of the laboratory period, be sure to keep a copy of your calculation for use while performing the experiment. The diluted NaOH solution will be used in the titration.

Next it will be necessary for you to calculate the quantity of solid oxalic acid dihydrate, $H_2C_2O_4 \cdot 2\,H_2O$, that will react with approximately 15 mL of the diluted NaOH solution. This is a solution stoichiometry problem in which the first step is to find the number of moles of NaOH in 15 mL of solution at the approximate concentration just determined. The volume of a solution times its molarity yields the number of moles:

$$\text{Volume (liter)} \times \text{molarity}\left(\frac{\text{moles}}{\text{liter}}\right) = \text{moles} \qquad (23.2)$$

The balance of the problem is set up and solved in the usual stoichiometry pattern. This is the second of the two preliminary calculations, and it corresponds to Performance Goal 23–2. This calculation appears as Question 2 in the Advance Study Assignment. Again, be sure to keep a copy for use while performing the experiment.

After carefully weighing out three samples of solid oxalic acid dihydrate, you will dissolve it and perform the titration described by Equation 23.1. From the mass of oxalic acid you will be able to determine the number of moles of acid present. From the equation you will determine the number of moles of NaOH required in the neutralization. You will then know both the volume of the NaOH solution and the number of moles of NaOH it contains. These are the "two essential items you require for (the) calculation of molarity" (see Introduction, p. 269): dividing moles by liters yields molarity. (See Performance Goal 23–3.)

The last of the performance goals for this experiment calls for you to perform titrations reproducibly. To meet this requirement you must come up with sodium hydroxide concentrations that are "the same" in separate titrations. This calls for establishing a standard of "sameness." You will be instructed to conduct three titrations as a minimum. If two of these yield molarities that are within 0.007 M of each other, they will be accepted as satisfying the reproducibility requirement. If you do not reach this result, additional titrations will be required.

So far nothing has been mentioned about the *accuracy* of your work. Indeed, within Experiment 23 there is no way to judge accuracy, as each student will have his/her own sodium hydroxide solution, which will have a concentration slightly different from that of his/her neighbor. In Experiment 24, however, you will use your standard solution to determine the concentration of an acid that is unknown to you, *but known to your instructor.* At this point an accurate result will be required. It should be apparent that your result in Experiment 24 cannot be accurate unless the concentration of the solution prepared in Experiment 23 has been determined accurately. As a consequence, if your accuracy in Experiment 24 does not meet the standard established, *it may be necessary for you to repeat Experiment 23 in order to correct previously undetected errors made there.* With this in mind, you are strongly urged to retain a complete record of *all* data, *all* volumes in the titrations, even if you think them to be incorrect. It is surprising how often "incorrect" data turn out to be just what is needed by the time a long experiment is completed.

SAMPLE CALCULATIONS

The following examples illustrate the calculations involved in Performance Goals 23–1 and 23–3:

Example 1

25.0 mL of a 12.0 M solution is diluted to 500 mL. Calculate the molarity of the dilute solution.

The number of moles of solute is the same in both the initial solution and the diluted solution; only water is added. This number of moles is

$$0.0250 \text{ L} \times \frac{12.0 \text{ moles}}{\text{L}} = 0.300 \text{ mole}$$

In the diluted solution the 0.300 mole of solute is dissolved in 500 mL, or 0.500 L. The concentration is therefore

$$\frac{0.300 \text{ mole}}{0.500 \text{ L}} = 0.600 \text{ mole/L}$$

Example 2

Calculate the molarity of an NaOH solution if a sample of oxalic acid weighing 1.235 g requires 42.5 mL of the base for neutralization.

First, determine the number of moles of oxalic acid present. The formula of solid oxalic acid is $H_2C_2O_4 \cdot 2\ H_2O$. Notice that, although the water of hydration is not shown in Equation 23.1, the acid is weighed as a solid, which includes the water. Accordingly, calculations must be based on the proper molar mass. Thus,

$$1.235 \text{ g } H_2C_2O_4 \cdot 2\,H_2O \times \frac{1 \text{ mole } H_2C_2O_4}{126 \text{ g } H_2C_2O_4 \cdot 2\,H_2O} = 0.00980 \text{ mole } H_2C_2O_4$$

Second, determine the number of moles of NaOH required to react with 0.00980 mole of $H_2C_2O_4$, according to the equation.

$$0.00980 \text{ mole } H_2C_2O_4 \times \frac{2 \text{ moles NaOH}}{1 \text{ mole } H_2C_2O_4} = 0.0196 \text{ mole NaOH}$$

Third, if 0.0196 mole of NaOH is present in 42.5 mL of solution, find the concentration in moles per *liter.*

$$\frac{0.0196 \text{ mole NaOH}}{0.0425 \text{ L}} = 0.461 \text{ M NaOH}$$

As a single dimensional analysis setup, this calculation would appear as:

$$1.235 \text{ g } H_2C_2O_4 \cdot 2\,H_2O \times \frac{1 \text{ mole } H_2C_2O_4}{126 \text{ g } H_2C_2O_4 \cdot 2\,H_2O} \times \frac{2 \text{ moles NaOH}}{1 \text{ mole } H_2C_2O_4} \times \frac{1}{0.0425 \text{ L}} = 0.461 \text{ M NaOH}$$

SAFETY PRECAUTIONS AND DISPOSAL METHODS

If some of the sodium hydroxide solution, either concentrated or dilute, comes into contact with your skin, it will have a slippery feeling, somewhat like soap. This is produced because the solution is slowly dissolving a layer of your skin. For obvious reasons, the process should not be allowed to continue. If you encounter that feeling at any time during the experiment, take time out to wash your hands thoroughly until the slippery feeling is gone.

Both the acid and base used in this experiment are corrosive and harmful over prolonged exposure. Avoid all unnecessary contact, and keep them off your clothing too. Both solutions are harmful to the eyes; be sure to wear goggles when working with all chemicals, either solid or in solution, during this experiment. The goggle requirement extends to cleaning-up operations, at which time bristles of buret brushes have been known to flick drops of chemicals into unprotected eyes.

After you have finished the titrations, SAVE your NaOH solution for Experiment 24. The content of the Erlenmeyer flasks can be poured down the drain.

PROCEDURE

Note: Record all mass measurements in grams to the nearest 0.001 g. Record all volume measurements from the buret in milliliters to the nearest 0.1 mL.

1. PREPARATION OF NaOH SOLUTION

A) Using a graduated cylinder, transfer about 100 mL of 1 M NaOH to a large beaker (600 mL or larger). With continuous stirring, dilute with deionized water until the total volume is about 500 mL.

Note: Thorough mixing is essential at this point. If you determine the "concentration" of an unmixed solution, you determine the concentration of only that part of the solution that you use. If you then use another part of the solution, with a different concentration, you will have no accuracy in your second application.

B) Transfer your solution to a stoppered or capped storage bottle. (Always keep your standard solution covered, as any evaporation loss or CO_2 absorption will change its concentration.) Label the bottle with your name, so it does not become lost among the bottles of your laboratory neighbors.

C) Calculate the approximate molarity of your solution. (This is Question 1 in the Advance Study Assignment.)

2. PREPARATION OF OXALIC ACID SOLUTIONS

A) From the approximate molarity of the diluted NaOH solution, calculate the mass of oxalic acid dihydrate, $H_2C_2O_4 \cdot 2\ H_2O$, needed to neutralize 15 mL of the base. Don't forget the water of hydration in the solid acid, as it will be present in what you weigh out. Have your instructor approve your calculation before proceeding. (This is Question 2 in the Advance Study Assignment.)

B) Make identifying marks on three 250-mL Erlenmeyer flasks.

Note: The next step is extremely critical. Your purpose is to transfer into each of the above Erlenmeyer flasks an amount of oxalic acid dihydrate that is approximately equal to what was calculated in Step 2A, but whose actual mass is known to the closest milligram. Be sure you understand that purpose and the procedure below by which it will be accomplished. The care with which this step is performed will determine both the accuracy and precision with which your sodium hydroxide solution is standardized, and the accuracy of your result in Experiment 24. Be particularly careful that no oxalic acid is spilled. If that happens, begin again.

C) Transfer four to five times the amount of oxalic acid dihydrate calculated in Step 2A into a small (preferably 50 mL or less), clean, dry beaker. Weigh the beaker and its contents on a milligram balance to the closest 0.001 g. Record the mass in the data table as "Mass of beaker + acid." Transfer from the beaker to one of the marked Erlenmeyer flasks a quantity of acid that is within 10 percent of the amount calculated in Step 2A by gently tapping the beaker. *Do not* use a spatula for the transfer or crystals will be lost on it. (Do not attempt to transfer exactly the calculated amount.) Weigh the beaker and its contents again, recording the measurement as "Mass of beaker − acid" for the first flask, and also as "Mass of beaker + acid" for the second flask. Again transfer the required amount of oxalic acid dihydrate, this time to the second Erlenmeyer flask. Weigh the beaker and its contents, recording the mass as "Mass of beaker − acid" for the second flask, and as "Mass of beaker + acid" for the third flask. Repeat the procedure once more to get the "Mass of beaker − acid" for the third flask.

D) Add about 50 mL of deionized water to each flask and swirl until the acid is dissolved. Be careful not to spill any water, which will take with it some of the weighed acid. (How important is the volume of water in which the acid is dissolved? How will the amount of water influence the number of milliliters of NaOH required to react with the quantity of acid in each flask?)

3. TITRATION OF OXALIC ACID WITH NaOH

A) Thoroughly clean your buret. Rinse the buret with about 10 mL of the NaOH solution. Drain and repeat with a second portion, discarding it as well. Fill the buret with the NaOH solution you prepared. Drain the solution into the calibrated part of the buret, letting the liquid flow through the tip. Make sure the tip of the buret is filled, and that there are no bubbles. Then record the initial buret reading to the nearest 0.1 mL.

B) Add 3 to 5 drops of phenolphthalein to each oxalic acid solution.

C) Hold one of the Erlenmeyer flasks so that the tip of the buret is inside the neck portion of the flask, as in Figure 23–1. Start the addition of NaOH. At the beginning of the titration you may add the base in larger portions, slowing down as the time required for the pink color to disappear gets longer. It is essential that you swirl the contents of the flask vigorously throughout the titration to assure complete mixing of the solutions. The end of the titration is reached when the pink color persists for 30 seconds. Record the buret reading at this point.

D) Repeat Step C with the other two samples. You may begin the titration from the volume reading on the buret at which the first titration ended. Be sure to record this value and make sure sufficient solution is available in the buret to complete the titration.

E) Calculate the results of your first three runs immediately. If you do not have two NaOH molarities within 0.007 of each other, repeat Parts 2 and 3 until reproducible results are obtained. *Keep all titration data!*

If you satisfy the reproducibility requirement, proceed to Experiment 24. If that experiment is to be done at another time, store your NaOH solution (see Step 1B). Clean all glassware thoroughly before putting it away or returning it to the stockroom.

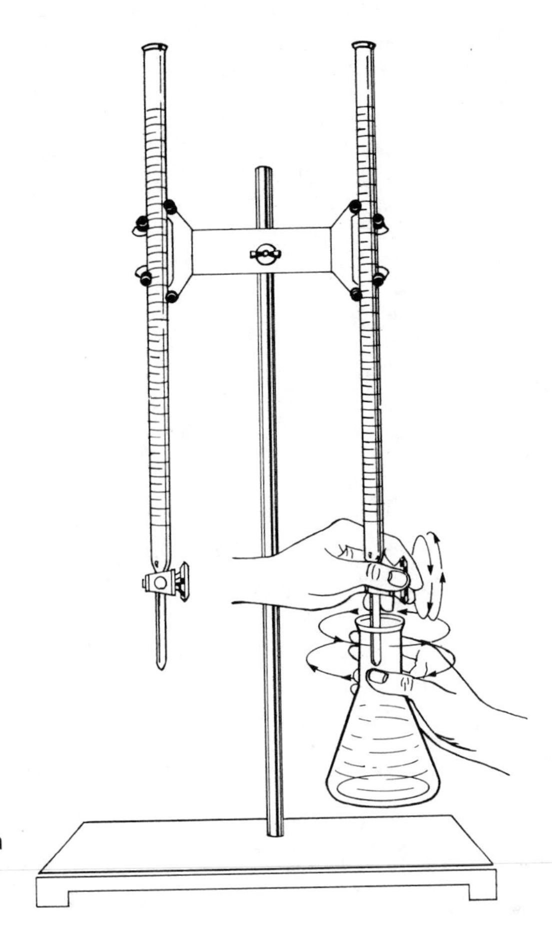

Figure 23–1. Titration from a buret into a flask.

Experiment 23
Work Page

NAME ....

DATE .... SECTION ....

TABLE OF DATA AND RESULTS (Beneath the table show the full calculation setup for at least one valid titration run)

Sample	*1*	*2*	*3*	*4*	*5*	*6*	*7*	*8*
Mass of beaker + acid (g)								
Mass of beaker − acid (g)								
Mass of oxalic acid (g)								
Initial buret reading (mL)								
Final buret reading (mL)								
Volume of NaOH (mL)								
Molarity of NaOH								

Calculation setup for at least one valid titration run:

Experiment 23

Report Sheet

NAME ....

DATE .... SECTION ....

TABLE OF DATA AND RESULTS (Beneath the table show the full calculation setup for at least one valid titration run)

Sample	*1*	*2*	*3*	*4*	*5*	*6*	*7*	*8*
Mass of beaker + acid (g)								
Mass of beaker − acid (g)								
Mass of oxalic acid (g)								
Initial buret reading (mL)								
Final buret reading (mL)								
Volume of NaOH (mL)								
Molarity of NaOH								

Calculation setup for at least one valid titration run:

Experiment 23

Advance Study Assignment

NAME ....

DATE .... SECTION ....

1) Read Procedure 1 for the preparation of your NaOH solution. Calculate the approximate molarity of the solution (Step 1C). Do not round off your result to the proper number of significant figures at this point.

2) Using your result from the problem above, calculate the mass of oxalic acid dihydrate, $H_2C_2O_4 \cdot 2\,H_2O$, required to react with 15 mL of your NaOH solution (Step 2A in the procedure). Round off your final answer, taking into account the information in Problem 1 as well as this one. It is this result that must be approved by your instructor before you go beyond Step 2A in the procedure.

3) If 0.628 g of $H_2C_2O_4 \cdot 2\,H_2O$ requires 24.7 mL of NaOH(aq) in a titration, calculate the molarity of the solution.

4) Why must you save your NaOH solution at the end of Experiment 23?

5) If your fingers feel slippery at any time during this experiment, what does that indicate and what should you do about it?

Experiment 24

Titration of Acids and Bases — II

Performance Goal

24–1 Determine the concentration of an acid by titrating a known volume with a standardized base.

CHEMICAL OVERVIEW

In this experiment you will use the standardized NaOH from Experiment 23 to determine the molar concentration of either hydrochloric acid (Option 1) or vinegar (Option 2), the latter being an acetic acid solution. The reactions for the two options are:

$$HCl(aq) + NaOH(aq) \rightarrow H_2O(\ell) + NaCl(aq) \quad \text{(Option 1)} \qquad (24.1)$$

$$HC_2H_3O_2(aq) + NaOH(aq) \rightarrow H_2O(\ell) + NaC_2H_3O_2(aq) \quad \text{(Option 2)} \qquad (24.2)$$

You will titrate a carefully measured volume of the acid with the NaOH solution. The product of the volume of NaOH times its molarity is the number of moles of base in the reaction. This is converted to moles of acid from the stoichiometry of the reaction. The number of moles of acid divided by the volume containing that number of moles (the volume of the acid sample) yields the molarity of the acid. These ideas were presented more fully in the general overview to Experiments 23 and 24.

To be acceptable, your reported molarity for the unknown must be within 0.015 of the molarity determined previously. This allows for an error of about 2 percent, which is quite large by analytical standards. Your results must also show good precision, that is, good reproducibility. In this experiment all sample sizes are the same. Therefore, all titration volumes should be alike. The requirement for precision will be satisfied if you have duplicate titrations within 0.2 mL (about 6 drops) of each other.

If your results are unsatisfactory from the standpoint of either precision or accuracy, your instructor may ask you to make repeat runs until acceptable results are reached. If precision is good but accuracy is bad, it is quite likely that the error lies in your results from Experiment 23. In this case your standardization runs against oxalic acid as a primary standard must be repeated. Keep all titration results, as they are helpful in locating errors if they do appear.

SAFETY PRECAUTIONS AND DISPOSAL METHODS

The same safety precautions identified for Experiment 23 apply to Experiment 24 — plus one more. In taking your samples of unknown acid, you will use a volumetric pipet. *Always* use a pipet bulb to draw liquid chemicals into a pipet; never use mouth suction.

After you have finished your titrations and checked your results with the instructor, the dilute NaOH can be poured down the drain and rinsed with plenty of cold water.

PROCEDURE

Note: Record all volume measurements in milliliters to the nearest 0.1 mL.

A) Clean and prepare your buret for use as in Experiment 23, Step 3A. Be sure to shake the NaOH solution before you start using it, especially if the solution was stored for a period of time. Fill the buret with the standardized NaOH solution from Experiment 23.

B) Pipet a 10.0-mL sample of your unknown acid — hydrochloric acid (Option 1) or vinegar (Option 2) — into each of three 250-mL Erlenmeyer flasks. Add about 50 mL of deionized water and 3 to 5 drops of phenolphthalein to each flask.

C) Titrate each acid sample, using the same procedure that you used in Step 3C of Experiment 23. If the three runs do not yield at least two volumes that are within 0.2 mL of each other, run three additional samples. Record and retain all data.

D) Calculate the molarity of your unknown acid and submit it for approval if your instructor so requests. If it is not approved, complete any additional titrations, either standardizations of the NaOH or analyses of the unknown, as required.

E) On receiving approval of your final result (if required by your instructor), thoroughly clean and rinse all glassware before putting it away or returning it to the stockroom.

Experiment 24

Work Page

NAME ....

DATE .... SECTION ....

TABLE OF DATA AND RESULTS (Beneath the table show the full calculation setup for at least one valid titration run)

Unknown Number ______________ Average Molarity of NaOH from Experiment 23 ______________

Sample	*1*	*2*	*3*	*4*	*5*	*6*	*7*	*8*
Initial buret reading (mL)								
Final buret reading (mL)								
Volume of NaOH (mL)								
Molarity of HCl (Option 1)								
Molarity of acetic acid (Option 2)								

Calculation setup for at least one valid titration run:

Experiment 24

Report Sheet

NAME ....

DATE .... SECTION ....

TABLE OF DATA AND RESULTS (Beneath the table show the full calculation setup for at least one valid titration run)

Unknown Number ______________ Average Molarity of NaOH from Experiment 23 ______________

Sample	*1*	*2*	*3*	*4*	*5*	*6*	*7*	*8*
Initial buret reading (mL)								
Final buret reading (mL)								
Volume of NaOH (mL)								
Molarity of HCl (Option 1)								
Molarity of acetic acid (Option 2)								

Calculation setup for at least one valid titration run:

Experiment 24

Advance Study Assignment

NAME ....

DATE .... SECTION ....

1) Why must you use a pipet bulb to draw a liquid into a pipet?

2) Why must you swirl the contents of the flask during the titration?

3) Give the calculation setup that shows that, if 15.0 mL of HCl(aq) requires 18.4 mL of 0.262 M NaOH in a titration, the molarity of the HCl is 0.321 M.

4) Calculate the molarity of an acetic acid solution, $HC_2H_3O_2$(aq), if a 25.0-mL sample requires 27.2 mL of 0.138 M NaOH in a titration.

Experiment 25

A Study of Reaction Rates

Performance Goal

25–1 State qualitatively and demonstrate experimentally the relationship between the rate of a chemical reaction and (a) temperature, (b) reactant concentration, and (c) the presence of a catalyst.

CHEMICAL OVERVIEW

The word "rate" implies change and time — change in some measurable quantity and the interval of time over which the change occurs. Speed in miles per hour, salary in dollars per month, and quantity in gallons per minute all express the idea of rate. The rate of a chemical reaction tells us how fast a reactant is being consumed or a product is being produced. Specifically, rate of reaction is the positive quantity that denotes how the concentration of a species in the reaction changes with time. For example, in the reaction

$$N_2(g) + 3\ H_2(g) \rightarrow 2\ NH_3(g) \qquad (25.1)$$

the rate of reaction can be expressed in terms of NH_3:

$$\text{Rate} = \frac{\text{change in concentration of } NH_3}{\text{time interval}} \qquad (25.2)$$

Alternatively, the rate could be expressed in terms of a reactant:

$$\text{Rate} = -\frac{\text{change in concentration of } N_2}{\text{time interval}} \qquad (25.3)$$

The minus sign is necessary to make the rate a positive quantity since the concentration of N_2 is decreasing with time.

The rate of a chemical reaction can be changed by (a) varying the concentration of reactants, (b) changing the temperature, or (c) introducing a catalyst.* In this experiment, you will study the effect of concentration by varying the concentration of one reactant while holding others constant; you will examine the rates of the same reaction at different temperatures; and you will conduct two reactions that will be identical in all respects except that a catalyst will be present in one and not in the other.

In analyzing the results of this experiment, *time* of reaction will be used as an indication of rate. Time and rate are inversely related: the higher the rate, the shorter the time. This is evident if you think of driving from one location to another: it takes less time if you drive at a higher rate (speed).

The principal reaction whose rate will be studied in this experiment is the iodine "clock reaction." It is the reaction between two solutions, solution A and solution B, that contain, among other things, an iodide and soluble starch. When the two solutions are combined, a series of reactions begins, ending with the release of elemental iodine. The appearance of iodine in the presence of starch may be detected easily by the appearance of a deep blue color. This signals the completion of the reaction.

* Catalysts are substances that drastically alter reaction rates without being used up in the reaction.

SAFETY PRECAUTIONS AND DISPOSAL METHODS

Sulfuric acid can cause severe burns when in contact with skin. If you should spill any of the 6 M acid on yourself, *immediately* wipe it off with a dry cloth or paper towel and then rinse it off with large amounts of water. Wear safety goggles when performing this experiment. Beware that sulfuric acid generates a lot of heat when it comes in contact with water.

Solutions obtained in this experiment can be poured down the drain.

PROCEDURE

1. COMPARISON OF REACTION RATES

Sodium Oxalate and Potassium Permanganate

A) Place about 15 mL of 0.1 M sodium oxalate into a large test tube and add 2 dropperfuls of 6 M sulfuric acid.

B) Place in a second test tube an amount of deionized water equal to the total volume of solution in the first test tube. This will be used as a control, or a color standard with which to compare the test tube in which the reaction is taking place.

C) Add 8 drops of 0.1 M potassium permanganate to the control test tube and mix by stirring. Place the test tube in a test-tube rack.

D) Recording the time to the nearest second, add 8 drops of 0.1 M potassium permanganate to the sodium oxalate solution and mix by stirring. Place the test tube in the rack, and observe the gradual color change compared to the control test tube. Record the time at which the reaction is complete, shown by the solution becoming colorless (not tan). Proceed to the next step while you are waiting.

The Iodine Clock Reaction

E) Using your graduated cylinder, pour 10 mL of solution A into a test tube. Rinse your graduated cylinder with deionized water and pour into it 10 mL of solution B.

F) Place a small beaker on a piece of white paper. Noting the time of mixing to the nearest second, pour the two solutions simultaneously into the beaker. Record the time to the nearest second when you observe the first sign of a reaction, shown by the appearance of a dark blue color.

2. EFFECT OF CONCENTRATION

To study the effect of concentration on the reaction rate, you will keep total volume, concentration of solution A, and temperature constant, and vary only the concentration of solution B.

A) Pour 10 mL of solution A into each of four test tubes, as you did into one test tube in Step 1E. Place the test tubes into a test-tube rack. Rinse your cylinder with deionized water.

B) Label four other test tubes 1 to 4 and place them into the test-tube rack. Using a 10-mL graduated cylinder, pour into each test tube the quantities of deionized water and solution B shown in the following table:

Test Tube Number	*Volume of Deionized Water (mL)*	*Volume of Solution B (mL)*	*Total Volume (mL)*
1	2	8	10
2	4	6	10
3	6	4	10
4	8	2	10
Part 1E	0	10	10

The undiluted solution B used in Steps 1E and 1F is added to the table for comparison. Note that the test tube from 1E has the highest concentration of solution B, test tube number 1 the next highest, and test tube number 4 the lowest. Note also that the total volume is the same in each case.

C) As in Step 1F, at some time 0 on a watch, pour the solution from test tube number 1 and the solution A from one of the four test tubes simultaneously into a small beaker on a piece of white paper. Record the number of seconds required for the reaction. Empty the beaker and rinse it with deionized water. Repeat the procedure with the remaining test-tube combinations, recording the elapsed time in each case.

3. EFFECT OF TEMPERATURE

A) Using a graduated cylinder, measure 10 mL of solution A into each of four test tubes. Rinse the cylinder, and then measure 10 mL of solution B into each of four other test tubes. Measure and record the temperature of solution A in one of the test tubes. (We will assume that solution B is at the same temperature, which is room temperature.) As before, empty one test tube of solution A and one test tube of solution B simultaneously into a small beaker. Record the time required for the reaction. (This is a repeat of Step 1F, except that this time you have recorded the temperature of the reactants.)

B) Place a test tube of solution A and a test tube of solution B into a 250-mL beaker containing tap water and a few chunks of ice. Immerse a thermometer into the test tube containing solution A. Be sure not to let the thermometer rest on the bottom of the test tube. (Again, we will assume that solution B, treated in the same manner as solution A, will have the same temperature.) Stir the contents of the beaker with the test tubes. When the temperature drops to about 10°C below room temperature, remove both test tubes from the ice bath and immediately pour the two solutions into a beaker, as before. Record both the temperature of the solutions and the time of the reaction.

C) Repeat the above procedure, replacing the ice-water bath with warm tap water. When the temperature of solution A has risen about 10°C above room temperature, pour the two solutions together. Obtain one more reading at about 20°C above room temperature. For each reading, record the solution temperature and the reaction time in the report sheet.

4. EFFECT OF A CATALYST

Measure 10 mL of solution A into a test tube and add a drop of 0.1 M $CuSO_4$ catalyst. In another test tube, combine 2 mL of solution B and 8 mL of deionized water (the same procedure as in Part 2, test tube number 4). Pour the solutions simultaneously into a beaker and record the time required for the reaction.

RESULTS

Part 2. Using the graph paper provided, plot a graph of milliliters of solution B divided by total milliliters versus reaction time. What relationship seems to be evident between concentration of solution B and the time needed for the blue color to appear?

Part 3. Using the graph paper provided, plot a graph of temperature versus reaction time.

Part 4. Compare the reaction times of the catalyzed reaction (Part 4) and the uncatalyzed reaction (Part 2, test tube 4) of solutions of the same concentration. How does the time required for the blue color to appear in this step compare with the time you observed for test tube number 4 in Part 2? Record your comparison on the report sheet.

Experiment 25
Work Page

NAME ....

DATE .... SECTION ....

PART 1 — COMPARISON OF REACTION RATES

	Sodium oxalate + Potassium permanganate	*Iodine Clock Reaction*
Time reaction began		
Time reaction completed		
Time of reaction (sec)		

Reaction having higher rate: Sodium oxalate + potassium permanganate ___________ ; iodine clock ___________ .

PART 2 — EFFECT OF CONCENTRATION ON THE IODINE CLOCK REACTION

Test Tube	$\frac{\textit{Volume of Solution B}}{\textit{Total Volume}}$	*Time (sec)*
1		
2		
3		
4		

Reaction rate varies directly ___________ , inversely ___________ with concentration of solution B.

PART 3 — EFFECT OF TEMPERATURE ON THE IODINE CLOCK REACTION

Temperature (°C)	*Time (sec)*

Increasing the temperature increases ___________ , decreases ___________ the reaction rate.

PART 4 — EFFECT OF CATALYST ON THE IODINE CLOCK REACTION

Reaction	*Time (sec)*
Without catalyst (test tube number 4, Part 2)	
With catalyst (from Part 4)	

Which *reaction rate* is greater? Catalyzed ___________ ; uncatalyzed ___________ .

Experiment 25

Work Page

NAME ....

DATE .... SECTION ....

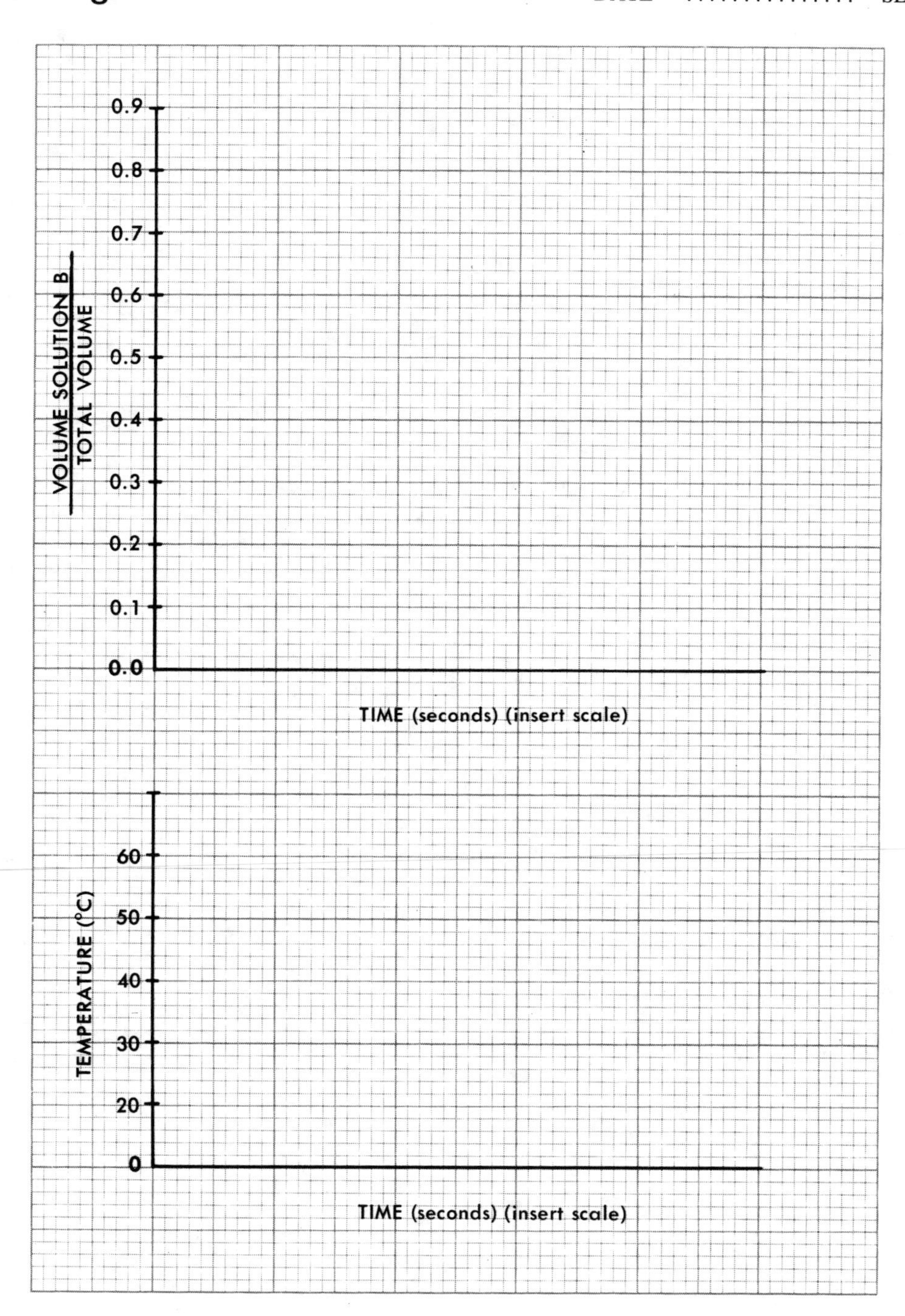

Experiment 25
Report Sheet

NAME ....

DATE .... SECTION ....

PART 1 — COMPARISON OF REACTION RATES

	Sodium oxalate + Potassium permanganate	*Iodine Clock Reaction*
Time reaction began		
Time reaction completed		
Time of reaction (sec)		

Reaction having higher rate: Sodium oxalate + potassium permanganate ____________ ; iodine clock ____________ .

PART 2 — EFFECT OF CONCENTRATION ON THE IODINE CLOCK REACTION

Test Tube	$\frac{\textit{Volume of Solution B}}{\textit{Total Volume}}$	*Time (sec)*
1		
2		
3		
4		

Reaction rate varies directly ____________ , inversely ____________ with concentration of solution B.

PART 3 — EFFECT OF TEMPERATURE ON THE IODINE CLOCK REACTION

Temperature (°C)	*Time (sec)*

Increasing the temperature increases ____________ , decreases ____________ the reaction rate.

PART 4 — EFFECT OF CATALYST ON THE IODINE CLOCK REACTION

Reaction	*Time (sec)*
Without catalyst (test tube number 4, Part 2)	
With catalyst (from Part 4)	

Which *reaction rate* is greater? Catalyzed ____________ ; uncatalyzed ____________ .

1) State the meaning of the term "reaction rate."
2) What is a catalyst?
3) If you wanted to alter the rate of a chemical reaction, what changes would you make in the experimental conditions?

NAME ....

DATE .... SECTION ....

Experiment 25

Report Sheet

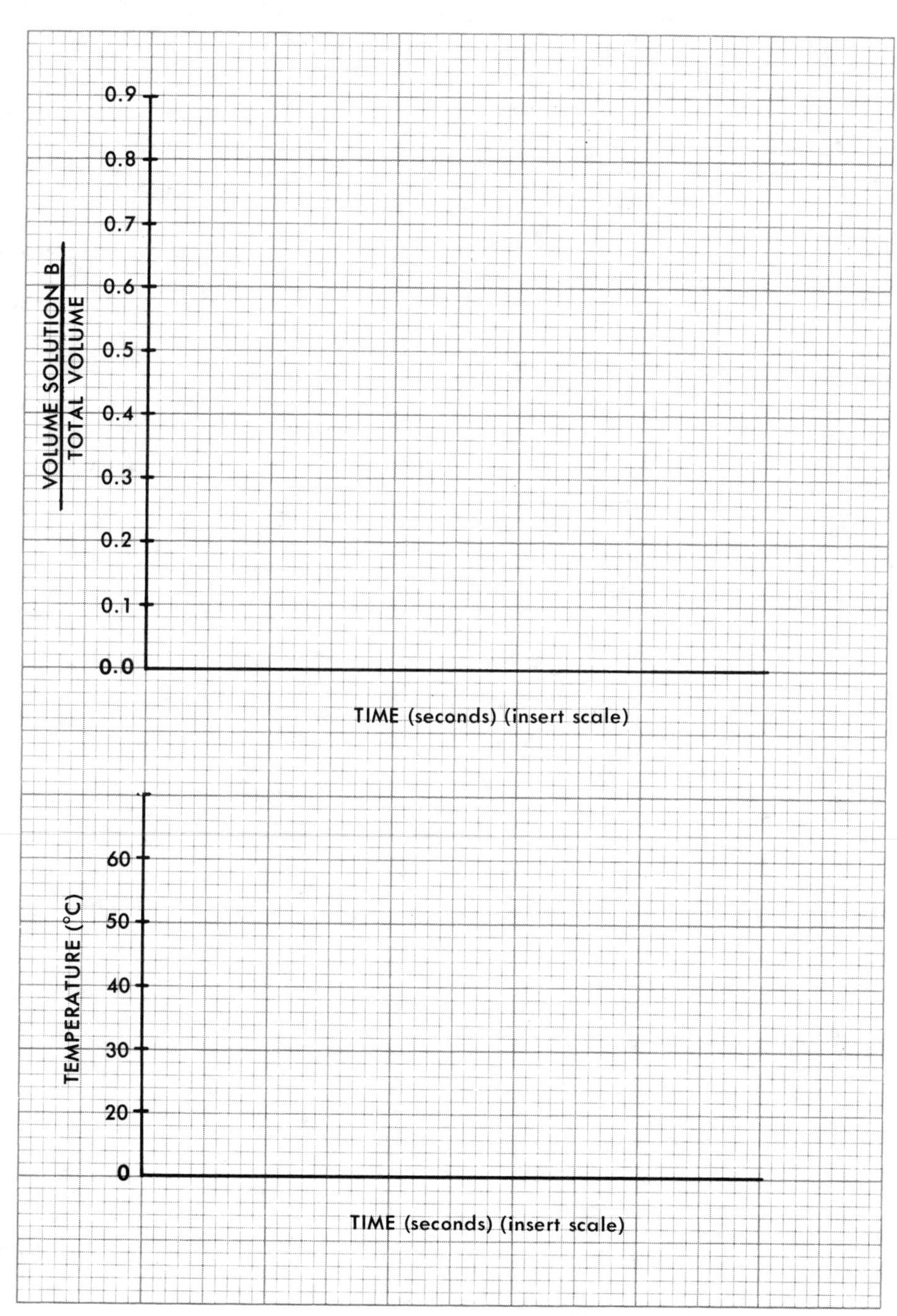

NAME ....

Experiment 25
Advance Study Assignment

DATE .... SECTION ....

1) State the meaning of the term "reaction rate."

2) What is a catalyst?

3) If you wanted to alter the rate of a chemical reaction, what changes would you make in the experimental conditions?

Experiment 26Ⓜ

Chemical Equilibrium

Performance Goal

26–1 Given the equation for a chemical equilibrium, predict and explain, on the basis of LeChatelier's Principle, the direction of a shift in the position of an equilibrium caused by a change in the concentration of one species.

CHEMICAL OVERVIEW

Some chemical reactions *proceed to completion,* until one of the reacting species is for all practical purposes completely consumed. One example of such a reaction is the precipitation of Cl^- ions from solution:

$$Ag^+(aq) + Cl^-(aq) \rightarrow AgCl(s) \tag{26.1}$$

When the Cl^- concentration is essentially zero, the reaction is complete.

Other reactions are *reversible.* This means that when the reactants are introduced into the reaction vessel, the reaction will start, but as soon as the reaction products begin to accumulate, they will react with each other to produce some of the starting species. At any time, *two* reactions are occurring, one going in the *"forward"* direction and one in *"reverse."* For example, consider the ionization of acetic acid in water:

$$CH_3COOH(aq) + H_2O(\ell) \rightleftarrows CH_3COO^-(aq) + H_3O^+(aq) \tag{26.2}$$

While some neutral acetic acid molecules are reacting with water, some of the hydronium ions (H_3O^+) and acetate ions (CH_3COO^-) formed by the ionization are recombining to yield the undissociated acid. The two arrows indicate the simultaneous occurrence of two reactions. *When the rate of the forward reaction exactly equals the rate of the reverse reaction, the system is said to be at equilibrium,* and no more detectable change occurs. This condition does not mean that all reactions have ceased, but only that the opposing reactions proceed at the same rate.

Consider the generalized reaction

$$A + B \rightleftarrows C + D \tag{26.3}$$

When the concentration of any one of the species in this equilibrium is changed, the equilibrium is "disturbed," and a "shift" will occur, either in the forward or reverse direction. **LeChatelier's Principle** predicts the direction of such a shift by stating: *When some stress is applied to a system originally at equilibrium, the system (reaction) will shift in such a direction as to counteract the stress, until a new equilibrium is reached.*

Let us consider how we can apply the preceding principle to the reaction shown in Equation 26.3. Suppose we add more of compound A. What will happen? The outside stress is an increase in the concentration of A. The reaction will shift in a direction that will counteract this increase. That is, the reaction will shift to reduce the concentration of A. A is consumed, and its concentration reduced, if the reaction shifts in the forward direction (as you read the equation from left to right). Based on the same reasoning, we can predict that if more C is added, the reverse shift will occur (consuming some of C). The displacement of an equilibrium by the addition of more of one of the species involved in the equilibrium is known as the **common ion effect.**

Evidence of a shift in equilibrium can easily be observed in the laboratory if one or more of the substances are colored or if a change in phase, such as precipitation or dissolution, accompanies the shift.

In this experiment, you will observe qualitatively the effect of changing the concentration of one or more substances in a chemical equilibrium, and then correlate your observations with LeChatelier's Principle. A chemical description of each equilibrium you will study is given in the Procedure section.

SAFETY PRECAUTIONS AND DISPOSAL METHODS

Solutions of sodium hydroxide react with your skin (or eyes), giving a slippery feeling. If this should occur, wash with plenty of water until no more slippery feeling is detected. Handle acid solutions with care, avoiding contact with skin. Stopper all reagent bottles as soon as you are through using them. Wear goggles when performing this experiment.

Solutions containing heavy metal ions (Co^{2+} and Fe^{3+}) should be disposed of in bottles provided. The rest of the solutions can be poured down the drain.

PROCEDURE

1. COBALT(II) ION COMPLEXES

Cobalt(II) ions, Co^{2+}, exist in water as aquo-complexes, $Co(H_2O)_6^{2+}$, that have a pink color. Other complexes exhibit different colors; the $CoCl_4^{2-}$ complex, for example, is blue. Depending on the relative concentration of chloride ions, the equilibrium shown in the following equation may be altered to yield a solution that is more blue or more pink:

$$\underset{\textit{Pink}}{Co(H_2O)_6^{2+}(aq)} + 4\,Cl^-(aq) \rightleftarrows \underset{\textit{Blue}}{CoCl_4^{2-}(aq)} + 6\,H_2O(\ell) \quad (26.4)$$

A) Pour about 10–15 drops of cobalt(II) chloride solution into each of three small test tubes. To the first test tube, add 10 drops of concentrated hydrochloric acid. Note the change, if any.

B) To the second test tube, add a small amount of solid ammonium chloride, and shake to make a saturated solution. Compare the color of the solution with that in the third test tube, which contains only the cobalt(II) chloride solution. Place both test tubes in a beaker containing boiling water and note the results. Cool both tubes under tap water. Tabulate and explain your observations.

2. IONIZATION OF ACETIC ACID

Consider the equilibrium

$$CH_3COOH(aq) + H_2O(\ell) \rightleftarrows CH_3COO^-(aq) + H_3O^+(aq) \quad (26.5)$$

Since no species in this reaction is colored, an auxiliary reagent is needed to help detect any shift in the equilibrium. You will use an indicator, methyl orange, for this purpose. In strongly acidic solutions (high H_3O^+ concentrations), methyl orange is red. A decrease in H_3O^+ concentration will cause a color change from red to yellow, with an intermediate color of orange.

A) Pour 10–15 drops of 0.1 M acetic acid into a test tube and add 1 or 2 drops of methyl orange. Place a few crystals of sodium acetate, CH_3COONa, in the solution and shake gently to dissolve them. Explain your observations.

B) Repeat the procedure in Part A but, instead of sodium acetate, add a few drops of 1 M sodium hydroxide. Record and explain your observations.

3. IONIZATION OF AQUEOUS AMMONIA

When ammonia gas is dissolved in water the following equilibrium is established:

$$NH_3(aq) + H_2O(\ell) \rightleftarrows NH_4{}^+(aq) + OH^-(aq) \qquad (26.6)$$

This solution is also called ammonium hydroxide because of the presence of ammonium and hydroxide ions. As with the equilibrium in Part 2, no species is colored. Therefore, we will use an indicator to signal the shifts in equilibrium. Phenolphthalein is colorless in slightly basic solutions, but it turns pink as the OH^- ion concentration increases.

A) Pour 10–15 drops 0.1 M ammonia into a test tube and add 1 or 2 drops of phenolphthalein. Note the color of the solution. Add solid ammonium chloride to the solution and shake gently to dissolve the crystals. Record and explain your observations.

B) Repeat the procedure in Part A, but instead of solid ammonium chloride, add 10 drops of 1 M zinc chloride. Record and explain your observations. (Hint: $Zn(OH)_2$ is quite insoluble.)

4. THE THIOCYANO–IRON(III) COMPLEX ION

The thiocyano–iron(III) complex ion can be formed from iron(III) ions (Fe^{3+}) and thiocyanate ions (SCN^-) according to the equation

$$Fe^{3+}(aq) + SCN^-(aq) \rightleftarrows Fe(SCN)^{2+}(aq) \qquad (26.7)$$

Tan *Blood Red*

Pour 10–15 drops of 0.1 M iron(III) nitrate and 10–15 drops of 0.1 M potassium thiocyanate into a 50-mL beaker. Dilute with 25 to 30 mL of deionized water to reduce the intensity of the deep red color. Pour 2- or 3-mL portions of this solution into each of three test tubes and proceed as follows:

a) Add about 10 drops of 0.1 M iron(III) nitrate solution to the contents of the first test tube.
b) To the second test tube, add 10 drops of 0.1 M potassium thiocyanate solution.
c) To the third test tube, add 4 to 5 drops of 10 percent sodium hydroxide.

Record and explain any changes you observed in parts (a) through (c). Hint: You may want to check the solubilities of some iron(III) compounds (see Table 21–1).

5. SATURATED SODIUM CHLORIDE EQUILIBRIUM

When a saturated solution of sodium chloride is in contact with undissolved solute, the following equilibrium exists:

$$NaCl(s) \rightleftarrows Na^+(aq) + Cl^-(aq) \qquad (26.8)$$

Pour 10–15 drops of saturated solution into a small test tube and add a few drops of concentrated hydrochloric acid. Note the result and give an explanation for it.

Experiment 26
Work Page

NAME ....

DATE .... SECTION ....

1) COBALT(II) ION COMPLEX EQUILIBRIUM

$$Co(H_2O)_6^{2+}(aq) + 4\ Cl^-(aq) \rightleftarrows CoCl_4^{2-}(aq) + 6\ H_2O(\ell)$$

A) *HCl addition:* Color change, if any: ______________________________

Direction of shift (foward, reverse, none): ______________________________

Explanation:

B) *NH_4Cl addition:* Colors of solutions in test tubes 2 and 3 at different temperatures:

	Test Tube 2	***Test Tube 3***
Room temperature		
Boiling water temperature		
After cooling		

Explanation:

2) IONIZATION EQUILIBRIUM OF ACETIC ACID

$$CH_3COOH(aq) + H_2O(\ell) \rightleftarrows CH_3COO^-(aq) + H_3O^+(aq)$$

A) *CH_3COONa addition:* Color change, if any: ______________________

Direction of shift (forward, reverse, none): ______________________

Explanation:

B) *NaOH addition:* Color change, if any: ______________________

Direction of shift (forward, reverse, none): ______________________

Explanation:

3) IONIZATION EQUILIBRIUM OF AQUEOUS AMMONIA

$$NH_3(aq) + H_2O(\ell) \rightleftarrows NH_4^+(aq) + OH^-(aq)$$

A) *NH_4Cl addition:* Color change, if any: ______________________

Direction of shift (forward, reverse, none): ______________________

Explanation:

NAME ....

Experiment 26
Work Page

DATE .... SECTION ....

B) *$ZnCl_2$ addition:* Color change, if any: ______________________

Direction of shift (forward, reverse, none): ______________________

Explanation:

4) THIOCYANO–IRON(III) COMPLEX ION EQUILIBRIUM

$$Fe^{3+}(aq) + SCN^{-}(aq) \rightleftarrows Fe(SCN)^{2+}(aq)$$

A) *$Fe(NO_3)_3$ addition:* Color change, if any: ______________________

Direction of shift (forward, reverse, none): ______________________

Explanation:

B) *KSCN addition:* Color change, if any: ______________________

Direction of shift (forward, reverse, none): ______________________

Explanation:

C) *NaOH addition:* Color change, if any: ______________________

Direction of shift (forward, reverse, none): ______________________

Explanation:

5) SATURATED SODIUM CHLORIDE EQUILIBRIUM

$$NaCl(s) \rightleftarrows Na^{+}(aq) + Cl^{-}(aq)$$

HCl addition: Change, if any: ______________________

Direction of shift (forward, reverse, none): ______________________

Explanation:

Experiment 26
Report Sheet

NAME ....

DATE .... SECTION ....

1) COBALT(II) ION COMPLEX EQUILIBRIUM

$$Co(H_2O)_6^{2+}(aq) + 4\ Cl^-(aq) \rightleftarrows CoCl_4^{2-}(aq) + 6\ H_2O(\ell)$$

A) *HCl addition:* Color change, if any: ______________________________

Direction of shift (foward, reverse, none): ______________________________

Explanation:

B) *NH_4Cl addition:* Colors of solutions in test tubes 2 and 3 at different temperatures:

	Test Tube 2	***Test Tube 3***
Room temperature		
Boiling water temperature		
After cooling		

Explanation:

2) IONIZATION EQUILIBRIUM OF ACETIC ACID

$$CH_3COOH(aq) + H_2O(\ell) \rightleftarrows CH_3COO^-(aq) + H_3O^+(aq)$$

A) *CH_3COONa addition:* Color change, if any: ______________________________

Direction of shift (forward, reverse, none): ______________________________

Explanation:

B) *NaOH addition:* Color change, if any: ______________________________

Direction of shift (forward, reverse, none): ______________________________

Explanation:

3) IONIZATION EQUILIBRIUM OF AQUEOUS AMMONIA

$$NH_3(aq) + H_2O(\ell) \rightleftarrows NH_4^+(aq) + OH^-(aq)$$

A) *NH_4Cl addition:* Color change, if any: ______________________________

Direction of shift (forward, reverse, none): ______________________________

Explanation:

Experiment 26
Report Sheet

NAME ....

DATE .... SECTION ....

B) *$ZnCl_2$ addition:* Color change, if any: ______________________________

Direction of shift (forward, reverse, none): ______________________________

Explanation:

4) THIOCYANO–IRON(III) COMPLEX ION EQUILIBRIUM

$$Fe^{3+}(aq) + SCN^{-}(aq) \rightleftarrows Fe(SCN)^{2+}(aq)$$

A) *$Fe(NO_3)_3$ addition:* Color change, if any: ______________________________

Direction of shift (forward, reverse, none): ______________________________

Explanation:

B) *KSCN addition:* Color change, if any: ______________________________

Direction of shift (forward, reverse, none): ______________________________

Explanation:

C) *NaOH addition:* Color change, if any: ______________________________

Direction of shift (forward, reverse, none): ______________________________

Explanation:

5) SATURATED SODIUM CHLORIDE EQUILIBRIUM

$$NaCl(s) \rightleftarrows Na^{+}(aq) + Cl^{-}(aq)$$

HCl addition: Change, if any: ______________________________

Direction of shift (forward, reverse, none): ______________________________

Explanation:

Experiment 26
Advance Study Assignment

NAME ....

DATE .... SECTION ....

1) State how and explain why the equilibrium $2X + Y \rightleftarrows Z$ will shift if
   a) X is removed

   b) Extra Y is added

   c) Some Z is added

2) Define, state, or describe:
   a) The common ion effect

   b) LeChatelier's Principle

   c) Are there any reactions occurring when a system is at equilibrium?

3) Consider the equilibrium $2\ CrO_4^{2-}(aq) + 2H^+(aq) \rightleftarrows Cr_2O_7^{2}(aq) + H_2O(\ell)$
Predict the direction the equilibrium will shift upon the

a) addition of NaOH

b) addition of hydrochloric acid

4) Give reasons for your predictions in Question 3.

Experiment 27

Measurement of pH with Indicators

Performance Goals

27–1 Prepare a set of pH indicator standards.
27–2 Measure the pH of an unknown solution by using indicators.

CHEMICAL OVERVIEW

Solutions of strong electrolytes such as strong acids and strong bases are good conductors of electricity. This indicates a high concentration of ions. In fact, strong acids and bases break into ions almost completely by either of two processes: **Dissociation** is the term used to describe the release of existing ions when an ionic compound dissolves, as in

$$NaOH(s) \rightarrow Na^{+}(aq) + OH^{-}(aq) \quad (27.1)$$

Ionization is the process whereby ions are formed when a covalent compound reacts with water, as in

$$HCl(g) + H_2O(\ell) \rightarrow Cl^{-}(aq) + H_3O^{+}(aq) \quad (27.2)$$

Even though the terms "ionization" and "dissociation" do not mean exactly the same thing, they are closely related and are often used interchangeably.

By contrast, solutions of weak electrolytes, such as weak acids and weak bases, are poor conductors of electricity. Since current is carried by mobile ions, this indicates a low concentration of ions. We therefore conclude that weak acids and bases are only partially ionized in water solutions. When acetic acid ionizes by reaction with water, equilibrium is reached:

$$CH_3COOH(aq) + H_2O(\ell) \rightleftarrows CH_3COO^{-}(aq) + H_3O^{+}(aq) \quad (27.3)$$

At equilibrium, acetic acid is only about 1 percent ionized, compared with HCl, which is nearly 100 percent ionized, as shown by Equation 27.2. Relatively few acetate (CH_3COO^-) and hydronium (H_3O^+) ions are present at equilibrium, but unionized acetic acid molecules (CH_3COOH) are in abundance.

The *acidity* of an aqueous solution is a measure of the concentration of the hydrogen (H^+) or hydronium (H_3O^+) ion.

> **Note: The hydronium ion may be considered a hydrated hydrogen ion, $H^+ \cdot H_2O$. The H^+ ion is easier to work with and will be used hereafter. It should be understood, however, that this ion is hydrated in aqueous solution and does not exist as a simple H^+.**

A convenient way to express the low acidity of weak acids is to use the **pH** scale. The pH of a solution is mathematically related to the hydrogen ion concentration by the equation

$$pH = -\log [H^+] = \frac{1}{\log [H^+]} \qquad (27.4)$$

where $[H^+]$ is the concentration of the hydrogen ion in moles per liter. By the mathematics of this equation, pH is the negative of the exponent of 10 that expresses the hydrogen ion concentration. For example, if pH = 5, then $[H^+] = 10^{-5}$; and a solution whose pH = 8 has a hydrogen ion concentration of 10^{-8}, or $[H^+] = 10^{-8}$.

Water ionizes into hydrogen and hydroxide ions:

$$H_2O(\ell) \rightleftarrows H^+(aq) + OH^-(aq) \qquad (27.5)$$

At 25°C, the ion product of water — the hydrogen ion concentration multiplied by the hydroxide ion concentration — is equal to 1.0×10^{-14}, or

$$[H^+]\,[OH^-] = 1.0 \times 10^{-14} \qquad (27.6)$$

If the ionization of water is the only source of these ions, it follows that they must be equal in concentration:

$$[H^+] = [OH^-] = 1.0 \times 10^{-7} \qquad (27.7)$$

A solution in which the hydrogen ion concentration is equal to the hydroxide ion concentration is said to be **neutral.** The pH of a neutral solution is, by calculation, 7. If the hydrogen ion concentration is greater than the hydroxide ion concentration, the solution is said to be **acidic.** In an acidic solution, the pH is less than 7. This is because the negative exponent of 10 decreases as the hydrogen ion concentration increases. $[H^+] = 10^{-5}$ (pH = 5) is a larger hydrogen ion concentration than $[H^+] = 10^{-7}$ (pH = 7). Conversely, in **basic** solutions, the concentration of the hydroxide ion exceeds the concentration of the hydrogen ion, and the pH will be greater than 7. Note at this point that as the hydrogen ion concentration goes down, the acidity decreases and the pH increases. The pH scale is illustrated in Figure 27–1.

Indicators are organic substances that impart to a solution a color that depends upon its pH. Ordinarily, the color will change gradually over a range of about two pH units. In this experiment, you will use two different indicators in a set of solutions of known pH. By comparing colors, you then estimate the pH of an unknown solution.

SAMPLE CALCULATIONS

Example 1

Calculate the pH of a 0.001 M HNO_3 solution.

Since nitric acid is a strong acid, we assume that it is completely ionized. Therefore,

$$[H^+] = 0.001, \text{ or } 10^{-3} \text{ mole/L} \qquad pH = 3$$

Example 2

Calculate the pH of a solution that contains 0.01 mole of hydrogen ion in 100 mL.

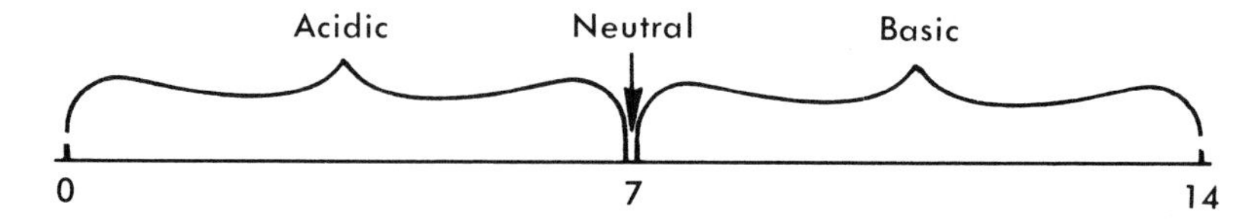

Figure 27–1. pH scale.

First, calculate $[H^+]$, always expressed in moles/liter:

$$\frac{0.01 \text{ mole H}^+}{0.100 \text{ L}} = 0.1 \text{ mole H}^+/\text{L} = 10^{-1} \text{ mole/L}$$

It follows that the pH is 1.

Example 3

Calculate the pH of a solution that is obtained when 25.0 mL of 2.0 M HNO_3 are diluted to 500 mL. Remember, nitric acid is assumed to be completely ionized. First, calculate the number of moles of H^+ present in 25.0 mL, or 0.0250 liter, of concentrated acid.

$$0.0250 \text{ L} \times \frac{2.0 \text{ moles H}^+}{\text{L}} = 0.050 \text{ mole H}^+$$

This many moles of hydrogen ions are present in the final volume, 500 mL. Therefore, the $[H^+]$ is

$$[\text{H}^+] = \frac{0.050 \text{ mole H}^+}{0.500 \text{ L}} = 0.10 \text{ mole/L} = 1.0 \times 10^{-1} \text{ mole/L}$$

The pH of the diluted solution is 1.00.

Alternately, you can carry out a dilution calculation in a single step by multiplying the original concentration by a "dilution factor," the original volume divided by the final volume:

$$[\text{H}^+] = \underbrace{\frac{2.0 \text{ moles H}^+}{\text{L (conc)}}}_{\textit{Original concentration}} \times \underbrace{\frac{0.0250 \text{ L (conc)}}{0.500 \text{ L (dil)}}}_{\textit{Dilution factor}} = \underbrace{0.10 \text{ mole H}^+/\text{L (dil)}}_{\textit{Final concentration}}$$

PROCEDURE

1. WATER PREPARATION

Since carbon dioxide from the air dissolves in water, yielding an acidic solution, we must remove all dissolved carbon dioxide from the water used in this experiment. Place 350 to 400 mL of deionized water in a beaker and heat it to boiling. Continue boiling for approximately 10 minutes, cover the vessel with a large watch glass, and allow it to cool to room temperature.

2. PREPARATION OF STANDARD SOLUTIONS

A) While the deionized water is being prepared, wash and label six test tubes having a capacity of greater than 10 milliliters. Label them 1 to 6. When the deionized water is at room temperature, prepare a set of solutions as described in Steps B to F. Figure 27–2 is a schematic diagram of the dilution procedure.

B) As accurately as possible, measure 5.0 mL of 1.0 M HCl into a 50-mL graduated cylinder. Your 5.0-mL measurement will be more accurate if you perform it in a separate 10-mL cylinder and then transfer the solution to the larger cylinder. Dilute the 5.0-mL HCl solution to 50.0 mL, again very accurately, with the treated deionized water. Transfer the contents to a small *dry* beaker and stir thoroughly with a glass rod. Pour 10.0 mL of this solution into test tube number 1.

C) Carefully and accurately measure 5.0 mL of the solution prepared in Step B into a dry (or deionized-water-rinsed) 50-mL graduated cylinder and again dilute carefully to 50.0 mL with boiled deionized water. Transfer to a clean, *dry* beaker, stir, and pour 10.0 mL into test tube number 2.

D) Repeat the procedure, using 5.0 mL from Step C, diluting to 50.0 mL. This time, pour 10.0-mL samples into *two* test tubes, number 3 and number 4.

E) Dilute 5.0 mL of solution from Step D to 50.0 mL in the same fashion and pour 10.0 mL into test tube number 5.

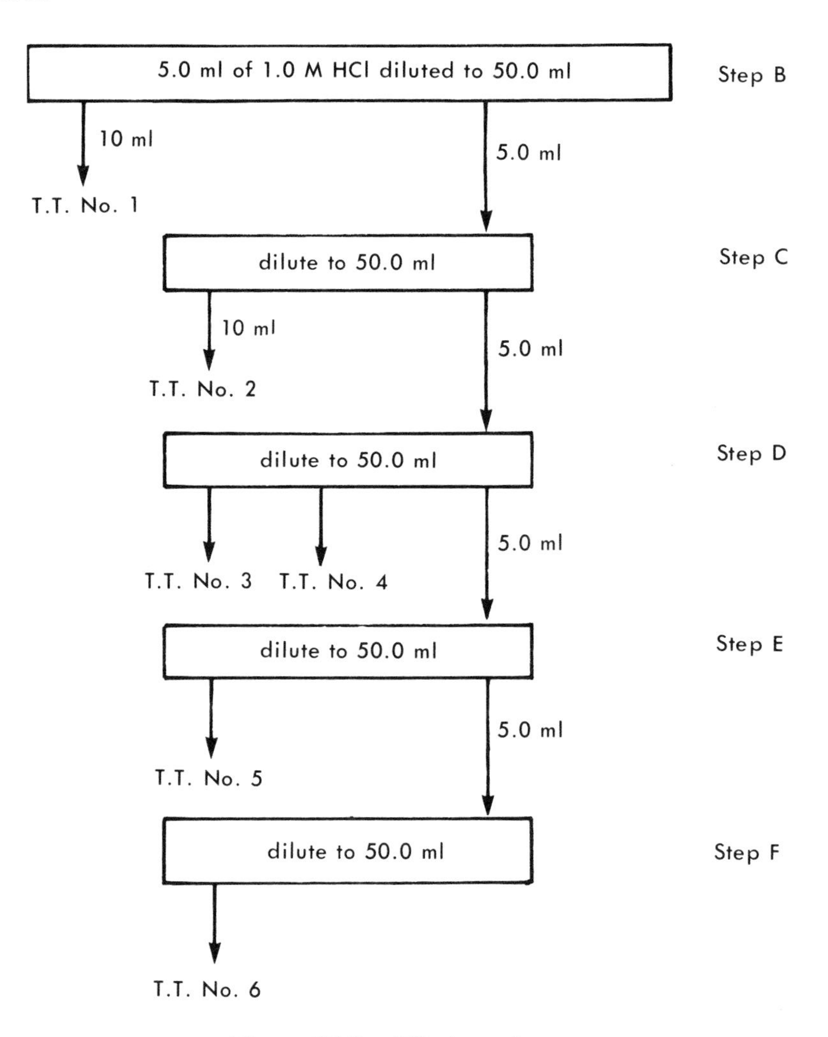

Figure 27–2. Dilution scheme.

F) Finally, dilute 5.0 mL from Step E to 50.0 mL and pour 10.0 mL into test tube number 6.

G) If you have not already done so, calculate the hydrogen ion concentration and, assuming complete ionization, the pH of each of the solutions you prepared. Enter the results in the report sheet.

H) To each of the test tubes numbered 1 through 3, add 2 drops of thymol blue indicator and mix well with a glass rod. *Be sure the rod is clean and dry before placing it in each solution.* Note the color in each test tube and record your observations in the report sheet. From these observations, estimate the pH range over which thymol blue changes color.

I) To test tubes 4 through 6, add 2 drops of methyl orange indicator and mix in the same fashion. Again record the observed color in each test tube, and estimate the pH range over which methyl orange changes color.

> **Note: If the methyl orange indicator solution is too concentrated, you will see the orange color of the indicator solution itself, regardless of what the pH is. Add only as much indicator solution as is necessary to give a clearly distinguishable color in the various test solutions, or dilute the indicator with deionized water and then add it to the test solutions.**

3. pH OF AN UNKNOWN SOLUTION

Obtain one or more unknowns from your instructor. Pour 10.0 mL of the first unknown into each of two test tubes. To the first test tube, add 2 drops of thymol blue indicator; to the second, add 2 drops of methyl orange (see note above). Mix thoroughly. Compare the colors of these solutions to your "standard" solutions. Basing your conclusions on the colors you observe, estimate and report the pH of the unknown solution. Repeat for each unknown.

Experiment 27
Work Page

NAME ....

DATE .... SECTION ....

Test Tube Number	*H^+ Concentration (moles/L)*	*pH*	*Indicator*	*Color Observed*
1				
2				
3				
4				
5				
6				
Unknown No. _____	X X X			
Unknown No. _____	X X X			
Unknown No. _____	X X X			

Estimated pH range of color transition:

a) Thymol blue: ______________________________

b) Methyl orange: ______________________________

Experiment 27

Report Sheet

NAME ....

DATE .... SECTION ....

Test Tube Number	*H^+ Concentration (moles/L)*	*pH*	*Indicator*	*Color Observed*
1				
2				
3				
4				
5				
6				
Unknown No. _____	X X X			
Unknown No. _____	X X X			
Unknown No. _____	X X X			

Estimated pH range of color transition:

a) Thymol blue: ________________________________

b) Methyl orange: ______________________________

Experiment 27

Advance Study Assignment

NAME ....

DATE .... SECTION ....

1) If a solution has a pH of 9.5, is it acidic or basic?

2) Calculate the pH of a 0.0001 M HCl solution. If you had an acetic acid solution of the same concentration, would its pH be higher or lower? Explain.

3) Steps 1B to 1F in the Procedure describe a series of dilutions, beginning with 5.0 mL of 1.0 M HCl. Step 1G asks you to calculate the hydrogen ion concentration and pH for each diluted solution. Perform these calculations now and enter the results in the table below.

Test Tube	*$[H^+]$*	*pH*	*Test Tube*	*$[H^+]$*	*pH*
1			4		
2			5		
3			6		

Experiment 28Ⓜ

Introduction to Oxidation–Reduction Reactions

Performance Goal

28–1 Determine experimentally the relative strengths of a selected group of oxidizing agents.

CHEMICAL OVERVIEW

Oxidation is defined as the process in which a loss of electrons occurs; **reduction** is a gain of electrons. From a broader viewpoint, in oxidation the oxidation number of an element increases (becomes more positive, as $+3 \rightarrow +5$, or $-3 \rightarrow -1$); whereas in reduction, the oxidation number decreases (becomes more negative, as $0 \rightarrow -1$, or $+7 \rightarrow +2$).

When a metal combines chemically with a halogen to form an ionic compound, an oxidation–reduction (redox) reaction occurs. Electrons are lost by the metal and gained by the halogen. Redox reactions may be thought of as electron transfer reactions, much as acid–base reactions may be viewed as proton transfer reactions. Each redox reaction may be considered the sum of two "half-reactions" or half-cell reactions:

$M \rightarrow M^{2+} + 2\,e^-$ Oxidation of a metal to M^{2+} ion by losing 2 electrons,

and

$X_2 + 2\,e^- \rightarrow 2\,X^-$ Reduction of a halogen to X^- by gaining 1 electron per atom.

Addition produces the net ionic redox equation:

$$\begin{array}{r} M \rightarrow M^{2+} + 2\,e^- \\ X_2 + 2\,e^- \rightarrow 2\,X^- \\ \hline M + X_2 \rightarrow M^{2+} + 2\,X^- \end{array}$$

Observe that the number of electrons lost by the metal exactly equals the number of electrons gained by the halogen. This balance is essential in any redox reaction; there can never be a deficiency or excess of electrons. As an ordinary chemical equation has to be balanced, so does a redox equation. Balancing is achieved by adjusting one or both half-reactions in order to equate the number of electrons lost and gained.

Example 1

$$\begin{array}{rl} A \rightarrow A^{2+} + 2\,e^- & \text{Oxidation} \\ B + e^- \rightarrow B^- & \text{Reduction} \end{array}$$

Multiply the reduction half-reaction by 2 and add:

$$\begin{array}{rl} A \rightarrow A^{2+} + 2\,e^- & \text{Oxidation} \\ 2\,B + 2\,e^- \rightarrow 2\,B^- & \text{Reduction} \\ \hline A + 2\,B \rightarrow A^{2+} + 2\,B^- & \text{Net ionic redox equation} \end{array}$$

Example 2

$$\begin{array}{rl} A \rightarrow A^{3+} + 3\,e^- & \text{Oxidation} \\ B + 2\,e^- \rightarrow B^{2-} & \text{Reduction} \end{array}$$

Multiply the oxidation reaction by 2 and the reduction reaction by 3 to equate the electrons lost in oxidation to the electrons gained in reduction:

$$\begin{array}{rl} 2\,A \rightarrow 2\,A^{3+} + 6\,e^- & \text{Oxidation} \\ 3\,B + 6\,e^- \rightarrow 3\,B^{2-} & \text{Reduction} \\ \hline 2\,A + 3\,B \rightarrow 2\,A^{3+} + 3\,B^{2-} & \text{Net ionic redox equation} \end{array}$$

In each of the preceding examples, the first element loses electrons to the second element; that is, the first element provides the electrons that reduce the second. Thus, the first element is referred to as a **reducing agent.** By accepting electrons, the second element causes the oxidation of the first element. Hence it is called an **oxidizing agent.** A summary of these terms is presented below:

If a species	*the species undergoes*	*and is called the*
Gains electrons	Reduction	Oxidizing agent (oxidizer)
Loses electrons	Oxidation	Reducing agent (reducer)

As acids vary in their strength (the ease with which they release protons), so reducers vary in their strength (the ease with which they release electrons). Similarly, oxidizers have different tendencies to capture electrons, just as bases vary in their attraction for protons. A strong oxidizer (oxidizing agent) has a great affinity for electrons.

In this experiment, you will investigate the relative ease with which certain metals and halides release electrons and will thereby build a partial qualitative chart of oxidizer strengths.

Note: Since metals and halides release electrons, they are the reducing agents, while metal ions and elemental halogens, Cl_2, Br_2, I_2, are the oxidizing agents.

Evidence of a metal–metal ion reaction is visible when a drop of solution containing the ion is placed on the metal. To detect a reaction between a halogen and a dissolved halide, however, a nonaqueous solvent that is immiscible with water must be used.

Halogens dissolved in trichloroethane have characteristic colors. Chlorine, Cl_2, in trichloroethane is colorless; bromine ranges from tan in dilute solutions to deep red or maroon when concentrated; and iodine is pale pink to deep purple, depending on concentration. You will use trichloroethane as a solvent to determine which halogen is present after a halogen and a halide ion have been combined. The trichloroethane is not involved in any chemical change; it serves simply as a solvent.

To interpret your observations correctly, consider the two examples below:

1. Suppose that the reaction between I_2 and Br^- goes to completion.

$$I_2 + 2\,Br^- \rightarrow 2\,I^- + Br_2$$

If iodine solution is added to trichloroethane, you will see a purple (pink) layer at the bottom of the test tube that is due to the presence of free I_2. After the addition of the KBr solution (the source of Br^- ions), the bottom layer's color will change to brown (tan/yellow), signaling the presence of Br_2.

2. Suppose that the reaction in the preceding example does not occur. Since there is no reaction, the purple color will remain unchanged, signifying that the original reactant, I_2, is still there.

SAFETY PRECAUTIONS AND DISPOSAL METHODS

Bromine in elemental form and concentrated iodine solution cause severe burns on contact with the skin. Also, bromine and chlorine waters release vapors that are extremely harmful when inhaled. These liquids should be handled only in a fume hood, and with utmost care. Trichloroethane vapors are harmful and should not be inhaled.

Do not pour solutions containing trichloroethane down the drain. Dispose of these solutions in a stoppered bottle.

PROCEDURE

1. METAL–METAL ION REACTIONS

A) Obtain one strip each of copper, zinc, and lead. Clean one side of each strip with emery paper. Lay the strips side by side on a paper towel on the desk, cleaned surface up.

B) Place on each strip 1 drop of each solution shown in Table 28–1. Record in your report sheet the combinations of metals and metal ions that showed evidence of a chemical change and those that did not. Wait about 5 minutes before you decide that no reaction has occurred. Include the silver ion, Ag^+, although the metal itself is not used because of its high cost. Assume that silver metal does not react with Zn^{2+}, Pb^{2+}, or Cu^{2+}.

C) On completing this part of the experiment, return or dispose of the metal strips as directed by your instructor.

2. HALOGENS

A) Arrange a series of six of your smallest test tubes in a rack and label them 1 through 6.

B) Pour about half an eyedropperful of trichloroethane into each test tube.

C) Add a few drops of chlorine water, bromine water, or iodine solution to the various test tubes as indicated in Table 28–2. Shake the solutions and record the initial color of each.

D) Add 8 to 10 drops of a halide solution to the test tubes, again following the directions given in Table 28–2. You may wish to do this part of the experiment one test tube at a time, to be sure not to get confused with the various reagents and colors. Shake the test tubes to mix the two solution layers. After shaking, allow the two phases to separate. Water and trichloroethane are immiscible, so trichloroethane, the denser liquid, will settle to the bottom, carrying any free halogen with it. By the color of this layer, determine which elemental halogen is present after mixing and thus whether or not a redox reaction has taken place. Has the initial color changed? If so, what does it mean? Record your observations.

Table 28–1 Solutions for Testing Metals

Metal	*Test Solutions*
Copper, Cu	Zn^{2+}, Pb^{2+}, Ag^+
Zinc, Zn	Cu^{2+}, Pb^{2+}, Ag^+
Lead, Pb	Cu^{2+}, Zn^{2+}, Ag^+

Table 28–2 Solution Combinations

Test Tube	*Halogen Solution*	*Halide Solution*
1	Cl_2	KBr
2	Cl_2	KI
3	Br_2	NaCl
4	Br_2	KI
5	I_2	NaCl
6	I_2	KBr

RESULTS

Record the answers to the following questions in the space provided in the report sheet:

1. Which combination(s), if any, yielded a redox reaction?
2. For each reaction that occurred, write the half-reaction equations for both oxidation and reduction.
3. If necessary, multiply either or both equations to equalize the electrons gained and lost in the half-reaction equations.
4. Add the half-reaction equations to get the net ionic redox equation.
5. From Part 1, list the oxidizers (oxidizing agents) in a column according to decreasing strength. Judge oxidizing strength by considering which species each oxidizer was capable of oxidizing and which species it could not oxidize. Show at the bottom the oxidizer that was incapable of oxidizing anything.
   Example: If metal A reacted with B^{2+} and C^{2+}, and metal B reacted with C^{2+} but not with A^{2+}, and metal C did not react with either A^{2+} or B^{2+}, then the strongest oxidizer is C^{2+} and the weakest oxidizer is A^{2+}. Remember, the oxidizers in this step are the metal ions.
6. Prepare a similar list for Part 2.

Experiment 28
Work Page

NAME ....

DATE .... SECTION ....

PART 1

Metal	Ion in Solution	Reaction	
		Yes	No
Cu	Zn^{2+}		
	Pb^{2+}		
	Ag^{+}		
Zn	Cu^{2+}		
	Pb^{2+}		
	Ag^{+}		
Pb	Cu^{2+}		
	Zn^{2+}		
	Ag^{+}		
Ag	Cu^{2+}		
	Zn^{2+}		
	Pb^{2+}		

Half-Reactions and Net Ionic Redox Equations

List of Oxidizers in
Order of Decreasing Strength

____

____

____

____

PART 2

Test Tube	*Halide Solution*	*Halide Ion*	*Halogen*	*Color*		*Reaction*	
				Initial	*After Mixing*	*Yes*	*No*
1	KBr	Br^-	Cl_2				
2	KI	I^-	Cl_2				
3	NaCl	Cl^-	Br_2				
4	KI	I^-	Br_2				
5	NaCl	Cl^-	I_2				
6	KBr	Br^-	I_2				

Half-Reactions and Net Ionic Redox Equations

List of Oxidizers in
Order of Decreasing Strength

____

____

____

Experiment 28
Report Sheet

NAME ....

DATE .... SECTION ....

PART 1

Metal	*Ion in Solution*	*Reaction* Yes	No
Cu	Zn^{2+}		
	Pb^{2+}		
	Ag^{+}		
Zn	Cu^{2+}		
	Pb^{2+}		
	Ag^{+}		
Pb	Cu^{2+}		
	Zn^{2+}		
	Ag^{+}		
Ag	Cu^{2+}		
	Zn^{2+}		
	Pb^{2+}		

Half-Reactions and Net Ionic Redox Equations

List of Oxidizers in
Order of Decreasing Strength

____

____

____

____

PART 2

Test Tube	*Halide Solution*	*Halide Ion*	*Halogen*	*Color*		*Reaction*	
				Initial	*After Mixing*	*Yes*	*No*
1	KBr	Br^-	Cl_2				
2	KI	I^-	Cl_2				
3	NaCl	Cl^-	Br_2				
4	KI	I^-	Br_2				
5	NaCl	Cl^-	I_2				
6	KBr	Br^-	I_2				

Half-Reactions and Net Ionic Redox Equations

List of Oxidizers in
Order of Decreasing Strength

____

____

____

NAME ....

Experiment 28

Advance Study Assignment

DATE .... SECTION ....

1) Define the following terms:

   a) Reduction

   b) Half reaction

   c) Oxidizer

2) Combine the following half-reactions to produce a balanced net ionic redox equation:

   a) $A \rightarrow A^{+} + e^{-}$

   $Z + 2\, e^{-} \rightarrow Z^{2-}$

   b) $B \rightarrow B^{2+} + 2\, e^{-}$

   $Y + 3\, e^{-} \rightarrow Y^{3-}$

   c) Which specie in reaction a) is the reducing agent?

Experiment 29

Hydrocarbons and Alcohols

Performance Goals

29–1 Perform various tests on an alkane to determine (a) whether its molar mass is relatively high or low, and (b) whether it is saturated or unsaturated.

29–2 Perform tests on an alcohol to determine if it is a primary, secondary, or tertiary alcohol.

CHEMICAL OVERVIEW

Organic chemistry is the study of the compounds of carbon other than the carbonates, cyanides, carbon monoxide, and carbon dioxide. All organic molecules contain carbon and hydrogen; many of them also contain oxygen. Organic compounds are divided into subgroups based on common "functional groups." These subgroups have similar chemical properties and can be identified by laboratory tests.

Hydrocarbons

The simplest types of organic compounds are the hydrocarbons, which are composed of carbon and hydrogen only. In these compounds, carbon atoms may join in an open chain or form ring structures:

```
    H   H   H                  H2
    |   |   |                  C
H — C — C — C — H            /   \
    |   |   |             H2C     CH
    H   H   H               |     ||
                          H2C     CH
                             \   /
                               C
                               H2
```

Propane (an alkane) *Cyclohexene (an alkene)*

Saturated hydrocarbons are those in which each carbon is bonded to four other atoms, either hydrogens or other carbon atoms. Hydrocarbons in this class are named **alkanes** or **paraffins.** The first few members of alkanes, such as methane and propane, are gases; the higher-molar-mass alkanes, such as those found in paraffin wax, are solids.

Unsaturated hydrocarbons contain one or more double or triple bonds. Those with double bonds are **alkenes** and those with triple bonds are **alkynes.** Addition to a double or triple bond characterizes these compounds:

$$\mathrm{>C{=}C<} + Br_2 \longrightarrow \mathrm{-\underset{Br}{\underset{|}{C}}-\underset{Br}{\underset{|}{C}}-} \qquad (29.1)$$

In general, alkenes and alkynes are much more reactive than alkanes are.

The reaction in Equation 29.1 is used to detect unsaturation. Bromine imparts a brownish color to a liquid. The disappearance of that color is a positive test for a multiple bond.

Unsaturated hydrocarbons can be oxidized by strong oxidizing agents, such as potassium permanganate, $KMnO_4$. Evidence of the reaction is the rapid disappearance of the purple color of the permanganate ion. Alcohols also react with $KMnO_4$.

Alcohols

Alcohols contain carbon, hydrogen, and oxygen, with the common functional group, —OH. This is known as the **hydroxyl group.** Depending on where the hydroxyl group is attached in a molecule, the alcohol is classified as primary (1°), secondary (2°), or tertiary (3°). In a primary alcohol the —OH group is connected to a carbon that is joined to only one other carbon atom. In a secondary alcohol the carbon that has the —OH group is joined to two other carbons, and in a tertiary alcohol the carbon is bonded to three other carbons.

$$R_1\text{—OH} \qquad R_1\text{—}\underset{\text{OH}}{\underset{|}{C}}\text{—}R_2 \qquad R_1\text{—}\overset{R_2}{\overset{|}{\underset{\text{OH}}{\underset{|}{C}}}}\text{—}R_3$$

Primary *Secondary* *Tertiary*

R_1, R_2, and R_3 represent alkyl groups in these diagrams. They may or may not be the same.

Alcohols are easily oxidized to either aldehydes or ketones:

$$R\text{—}\overset{H}{\overset{|}{\underset{\text{OH}}{\underset{|}{C}}}}\text{—}H \xrightarrow{\text{ox.}} R\text{—}\underset{O}{\underset{\|}{C}}\text{—}H \quad \text{(an aldehyde)} \tag{29.2}$$

$$R_1\text{—}\overset{H}{\overset{|}{\underset{\text{OH}}{\underset{|}{C}}}}\text{—}R_2 \xrightarrow{\text{ox.}} R_1\text{—}\underset{O}{\underset{\|}{C}}\text{—}R_2 \quad \text{(a ketone)} \tag{29.3}$$

Tertiary alcohols resist oxidation, even by strong oxidizing agents.

By oxidation and subsequent testing of the oxidation product, an alcohol can be classified as primary, secondary, or tertiary. The Lucas test will be used in this experiment to classify alcohols. Tertiary alcohols are converted to alkyl halides very rapidly, secondary alcohols take a few minutes, while primary alcohols react very slowly, with no observable change in 10 minutes.

$$R_1\text{—}\overset{R_2}{\overset{|}{\underset{\text{OH}}{\underset{|}{C}}}}\text{—}R_3 + \text{HCl} \xrightarrow{ZnCl_2} R_1\text{—}\overset{R_2}{\overset{|}{\underset{\text{Cl}}{\underset{|}{C}}}}\text{—}R_3 + H_2O \tag{29.4}$$

(Insoluble in water)

Because the —OH group imparts polar characteristics to alcohols, the lower-molar-mass alcohols are water soluble. As the length of the carbon chain increases, solubility decreases rapidly.

SAFETY PRECAUTIONS AND DISPOSAL METHODS

The organic chemicals you will use in this experiment are highly volatile and flammable. Do not breathe the vapors and be sure not to use them near flames. Bromine, Br_2, causes severe burns on contact with the skin. Handle this reagent very carefully and only in the fume hood. Avoid breathing the vapors, which are also very harmful. Be sure to wear eye protection while performing this experiment, and wash your hands when you are finished.

Dispose of the organic liquids in a special container provided or as directed by your instructor. *Do not pour organic liquids down the sinks in the open laboratory!*

PROCEDURE

1. HYDROCARBONS

A) **Solubility.** Add 5 mL of absolute ethanol to three clean, dry, labeled test tubes. Now pour 1 mL of each of the following liquid alkanes into separate test tubes and shake. Observe if the substance dissolves and if so, how rapidly. Record your observations in the report sheet.

Test tube No. 1: hexane (C_6H_{14})
Test tube No. 2: gasoline (assume to be C_8H_{18})
Test tube No. 3: kerosene (assume to be $C_{14}H_{30}$)

B) **Reaction with Br_2.** Add 1 mL of hexane (an alkane) and cyclohexene (an alkene) to two separate, dry, labeled test tubes. Add 2 to 3 drops of bromine solution to each and shake. Observe any color change and record it in the report sheet.

C) **Reaction with $KMnO_4$.** Add 1 mL of hexane and cyclohexene to two separate, dry, labeled test tubes. Add 2 to 3 drops of $KMnO_4$ solution to each and shake. The disappearance of the purple color of $KMnO_4$ with the appearance of a brown precipitate (MnO_2) is a positive test for unsaturated hydrocarbons.

D) **Test on an unknown hydrocarbon.** Obtain an unknown from your instructor. Pour 1-mL portions of it into three test tubes and perform tests A through C, above. Record your results on the report sheet. By comparing the results obtained on the knowns and your unknown, classify it as low- or high-molar-mass, saturated or unsaturated, hydrocarbon.

2. ALCOHOLS

A) **Lucas test.** Add 1 mL of ethanol, 2-propanol, and t-butyl alcohol to three separate, dry, labeled test tubes. Pour 3 mL of Lucas reagent into each, stopper, and shake to mix. Allow the solutions to stand at room temperature for up to 15 minutes, noting the time when a change occurs. The appearance of cloudiness is indication of a reaction. If no change is visible, immerse the test tube(s) into a water bath maintained at approximately 60°C and observe whether any change occurs. Discontinue heating after 15 minutes. Record your observations in the report sheet.

B) **Classification of an unknown.** Obtain an unknown from your instructor. Pour 1-mL of it into a dry test tube and perform the Lucas test. Record your observations and classify the unknown as primary, secondary, or tertiary alcohol.

Experiment 29
Work Page

NAME ....

DATE .... SECTION ....

1) HYDROCARBONS

A) Solubility

Test Tube	*Substance*	*Solubility*		
		Rapid	*Slow*	*None*
1				
2				
3				

B) Reaction with Br_2

Hexane reacts ____________ , does not react ____________ with Br_2.

Cyclohexene reacts ____________ , does not react ____________ with Br_2.

C) Reaction with $KMnO_4$

Substance	*Yes*	*No*
Hexane		
Cyclohexene		

If the structure of acetylene is HC≡CH, would you predict that it gives a positive ____________ or negative ____________ test with $KMnO_4$?

D) Unknown

Soluble in ethanol? Yes ______ No ______

Reaction with Br_2? Yes ______ No ______

Reaction with $KMnO_4$? Yes ______ No ______

Molar mass is low _____________ , high _____________ .

The compound is saturated _____________ , unsaturated _____________ .

2) ALCOHOLS

A) Lucas Test

Test Tube	*Substance*	*Lucas Test*		*Time (min)*	
		Yes	*No*	*Room Temp.*	*60°C*
1					
2					
3					

B) Unknown

Unknown Number______________________

If Lucas test positive at room temperature _____________ , 60°C _____________ , time for reaction _____________ minutes.

Based on the above observations, the unknown is a 1° _____________ , 2° _____________ , 3° _____________ alcohol.

Experiment 29

Report Sheet

NAME ....

DATE .... SECTION ....

1) HYDROCARBONS

A) Solubility

Test Tube	*Substance*	*Solubility*		
		Rapid	*Slow*	*None*
1				
2				
3				

B) Reaction with Br_2

Hexane reacts ____________ , does not react ____________ with Br_2.

Cyclohexene reacts ____________ , does not react ____________ with Br_2.

C) Reaction with $KMnO_4$

Substance	*Yes*	*No*
Hexane		
Cyclohexene		

If the structure of acetylene is HC≡CH, would you predict that it gives a positive ____________ or negative ____________ test with $KMnO_4$?

D) Unknown

Soluble in ethanol? Yes ______ No ______

Reaction with Br_2? Yes ______ No ______

Reaction with $KMnO_4$? Yes ______ No ______

Molar mass is low _____________ , high _____________ .

The compound is saturated _____________ , unsaturated _____________ .

2) ALCOHOLS

A) Lucas Test

Test Tube	*Substance*	*Lucas Test*		*Time (min)*	
		Yes	*No*	*Room Temp.*	*60°C*
1					
2					
3					

B) Unknown

Unknown Number______________________

If Lucas test positive at room temperature _____________ , 60°C _____________ , time for reaction _____________ minutes.

Based on the above observations, the unknown is a 1° _____________ , 2° _____________ , 3° _____________ alcohol.

Experiment 29

Advance Study Assignment

NAME ....

DATE .... SECTION ....

1) Describe the difference between the structures of alkanes and alkynes.

2) If the polarity of alcohols decreases with increasing molar mass, predict how the solubility of alcohols in water (a polar solvent) changes with increasing molar mass. Explain.

3) Complete the table below using the information given in the Chemical Overview:

Type of Alcohol	*Lucas Test*	
	Room Temp.	*60°C*
1°		
2°		
3°		

Experiment 30

Aldehydes, Ketones, and Carboxylic Acids

Performance Goal

30–1 Identify an unknown as an aldehyde, ketone, or carboxylic acid.

CHEMICAL OVERVIEW

Organic compounds are best studied according to the functional groups they contain. Aldehydes and ketones both contain the carbonyl group,

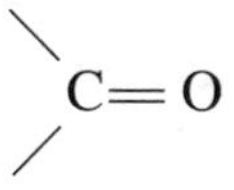

$$\rangle C{=}O$$

In aldehydes, one of the bonds is always connected to a hydrogen atom. Thus all aldehydes have a $\rangle C{=}O$ group. All carboxylic acids contain the carboxyl group,

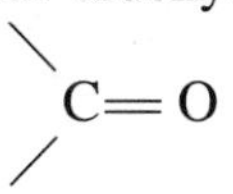

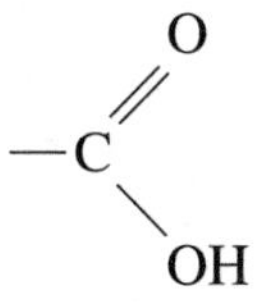

$$-C\langle^{O}_{OH}$$

Aldehydes and Ketones

When a primary (1°) alcohol is oxidized, the product is an aldehyde, while a secondary (2°) alcohol yields a ketone. The —CHO group of an aldehyde is easily oxidized to a carboxylic acid:

$$\underset{\textit{Aldehyde}}{R{-}CHO} \xrightarrow{\text{ox.}} \underset{\textit{Carboxylic acid}}{R{-}COOH} \qquad (30.1)$$

Ketones, on the other hand, resist oxidation. This difference is the basis of several common laboratory tests that distinguish aldehydes from ketones. Two such tests use metal ions as the oxidizing agent:

$$R{-}CHO \xrightarrow{Ag^+} Ag\ (s) + R{-}COOH \qquad \text{Tollen's test} \qquad (30.2)$$

$$R{-}CHO \xrightarrow{Cu^{2+}} Cu_2O\ (s) + R{-}COOH \qquad \text{Benedict's test or Fehling's test} \qquad (30.3)$$

A very common way to identify a ketone with an $R-\overset{\overset{O}{\|}}{C}-CH_3$ group is the iodoform test. Iodoform, CHI_3, is a yellow solid with a melting point of 119°C. The ketone reacts with I_2 in an alkaline solution to give CHI_3:

$$R-\overset{\overset{O}{\|}}{C}-CH_3 \xrightarrow{I_2,\ NaOH} R-COO^- + CHI_3(s) \qquad (30.4)$$

Iodoform has a characteristic medicine-like smell, which helps to establish its presence.

The lower-molar-mass aldehydes and ketones are water soluble, but as the carbon chain lengthens they become insoluble.

Carboxylic Acids

Most organic acids are weak acids because they ionize only slightly in aqueous solutions. For example, 1 molar acetic acid is only about 1 percent ionized; 99 percent of the compound remains in molecular form. Even though the hydrogen ion concentration is lower than it would be for the same concentration of a strong acid, such as HCl, HNO_3, or H_2SO_4, the solutions of organic acids *are* acidic. They have low pH values, whereas aldehydes and ketones are neutral.

Organic acids, except for those with low molar mass, are not soluble in water. Upon reaction with a strong base (NaOH), they form salts that *are* water soluble.

$$R-C\begin{matrix}\nearrow\!\!\!\!\!\!/\ O \\ \searrow OH\end{matrix} + OH^- \longrightarrow R-C\begin{matrix}\nearrow\!\!\!\!\!\!/\ O \\ \searrow O^-\end{matrix} + H_2O \qquad (30.5)$$

A typical acid–base reaction occurs between carboxylic acids and aqueous $NaHCO_3$. Gaseous CO_2 is produced according to

$$R-C\begin{matrix}\nearrow\!\!\!\!\!\!/\ O \\ \searrow OH\end{matrix} + HCO_3^- \longrightarrow R-C\begin{matrix}\nearrow\!\!\!\!\!\!/\ O \\ \searrow O^-\end{matrix} + H_2O + CO_2\,(g) \qquad (30.6)$$

Neither aldehydes nor ketones react with aqueous $NaHCO_3$.

When a carboxylic acid reacts with an alcohol, the product is called an ester (see Experiment 31). Many esters have distinct odors (oil of wintergreen, banana oil, apricot oil, etc.) and are widely used as scents or flavorings.

SAFETY PRECAUTIONS AND DISPOSAL METHODS

Several organic compounds used in this experiment are volatile and have very strong odors. Handle these chemicals in the hood and be careful not to get them on your skin. Acetone is flammable; do not use it near flames. Avoid breathing the vapors of chemicals; some are potentially harmful. Be sure to wear eye protection while performing this experiment.

Dispose of the organic liquids in a special container provided or as directed by your instructor. *Do not pour organic liquids down the sinks in the open laboratory!*

PROCEDURE

1. ALDEHYDES AND KETONES

A) **Benedict's test.** Label three clean, dry test tubes and pour 1 mL of formaldehyde solution, propanal, and acetone into separate test tubes. Prepare a fourth tube with deionized (distilled) water (to be used as a blank). Add to each test tube 5 mL of Benedict's reagent. Place the test tubes in a boiling-water bath for 2 to 3 minutes. The formation of a reddish-brown precipitate (Cu_2O) indicates the presence of an aldehyde. To be considered a positive test, a precipitate *must* be present. Enter your observations in the report sheet.

B) **Iodoform test.** Label three large test tubes and add 2 mL of water to each. Now add 5 drops of formaldehyde solution, propanal, and acetone to separate test tubes. Pour 1 mL of 10% NaOH solution into each and add dropwise, with shaking, enough iodine test solution to give a definite, dark iodine color. If at this point less than 2 mL (about 40 drops) of iodine solution was used, heat the test tube in a water bath at about 60°C. If the solution is decolorized on heating, add more iodine test solution until the dark color persists for 2 minutes at 60°C. Now add a few drops of 10% NaOH while shaking to expel any unreacted iodine. The solutions at this point will remain yellow.

Fill each test tube with deionized water and allow it to stand for 15 minutes. (Proceed with the next part of the experiment while waiting.) If a methyl ketone was present, a yellow precipitate (CHI_3) will form. Record your observations in the report sheet.

2. CARBOXYLIC ACIDS

A) **pH test.** Pour 5 mL of deionized water into four labeled test tubes. Add 1 mL of acetic acid to the first test tube, 1 mL of propionic acid (also called propanoic acid) to the second, and 0.1 g of salicylic acid to the third. The fourth test tube will serve as a blank. If the acid does not dissolve, use ethanol as a solvent instead of water and then add water until crystals just begin to form. Test the pH of each solution with a pH paper and record your observations in the report sheet.

B) **Sodium hydrogen carbonate test.** To three labeled test tubes add 3 mL of 10% sodium hydrogen carbonate ($NaHCO_3$) solution. To the first one add 10 drops of acetic acid; to the second, 10 drops of propionic acid; and to the third, 0.1 g of salicylic acid. A distinct fizzing sound or visual observation of the formation of CO_2 bubbles constitutes a positive test.

3. IDENTIFICATION OF AN UNKNOWN

Obtain an unknown from your instructor. If it is a solid, use 0.1-g portions; if it is a liquid, use the same amount you used for the knowns. Repeat tests A and B for aldehydes and ketones, followed by the tests for carboxylic acids. Based on your observations, classify your unknown as an aldehyde, ketone, or acid. Complete the report sheet.

Experiment 30
Work Page

NAME ....

DATE .... SECTION ....

1) ALDEHYDES AND KETONES

A) Benedict's Test

Test Tube	*Substance*	*Yes*	*No*
1	Formaldehyde		
2	Propanal		
3	Acetone		
4	H_2O		

B) Iodoform Test

Test Tube	*Substance*	*Room Temp.*		*60°C*	
		Yes	*No*	*Yes*	*No*
1	Formaldehyde				
2	Propanal				
3	Acetone				

2) CARBOXYLIC ACIDS

A) pH Test

Compare results obtained on the three knowns to the pH of your deionized (distilled) water.

Test Tube	*Substance*	*Approx. pH*
1	Acetic acid	
2	Propionic acid	
3	Salicylic acid	

Are the solutions acidic _________ , neutral _________ , or basic _________ ?

B) Sodium Hydrogen Carbonate Test

Test Tube	*Substance*	*Yes*	*No*
1	Acetic acid		
2	Propionic acid		
3	Salicylic acid		

C) Unknown

Unknown Number ________

Benedict's test: Yes ________ No ________

Iodoform test: Yes ________ No ________

pH test: Acidic ________ Neutral ________ Basic ________

Sodium hydrogen carbonate test: Yes ________ No ________

Compound is an aldehyde ________ , a ketone ________ , an acid ________ .

Experiment 30
Report Sheet

NAME ....

DATE .... SECTION ....

1) ALDEHYDES AND KETONES

A) Benedict's Test

Test Tube	*Substance*	*Yes*	*No*
1	Formaldehyde		
2	Propanal		
3	Acetone		
4	H_2O		

B) Iodoform Test

Test Tube	*Substance*	*Room Temp.*		*60°C*	
		Yes	*No*	*Yes*	*No*
1	Formaldehyde				
2	Propanal				
3	Acetone				

2) CARBOXYLIC ACIDS

A) pH Test

Compare results obtained on the three knowns to the pH of your deionized (distilled) water.

Test Tube	*Substance*	*Approx. pH*
1	Acetic acid	
2	Propionic acid	
3	Salicylic acid	

Are the solutions acidic _________ , neutral _________ , or basic _________ ?

B) Sodium Hydrogen Carbonate Test

Test Tube	*Substance*	*Yes*	*No*
1	Acetic acid		
2	Propionic acid		
3	Salicylic acid		

C) Unknown

Unknown Number _________

Benedict's test: Yes _________ No _________

Iodoform test: Yes _________ No _________

pH test: Acidic _________ Neutral _________ Basic _________

Sodium hydrogen carbonate test: Yes _________ No _________

Compound is an aldehyde _________ , a ketone _________ , an acid _________ .

Experiment 30

Advance Study Assignment

NAME ....

DATE .... SECTION ....

1) Answer the following questions as true (T) or false (F):

a) Organic acids are weak acids ____________

b) The pH of 0.10 M HCl is higher that the pH of 0.10 M acetic acid ____________

c) Sodium salts of organic acids are soluble in water ____________

d) Both aldehydes and ketones contain the carbonyl group ____________

2) Describe the difference in behavior between ketones and carboxylic acids when reacting with sodium hydrogen carbonate.

3) If an unknown compound in this experiment gave a negative Benedict's test, a positive iodoform test, and a negative sodium hydrogen carbonate test, it would be a(n) ____________________ (aldehyde, ketone, or carboxylic acid).

Experiment 31

Preparation of Aspirin

Performance Goal

31–1 Beginning with salicylic acid and acetic anhydride, prepare a sample of aspirin.

CHEMICAL OVERVIEW

Chemically speaking, aspirin is an organic ester. An ester is a compound that is formed when an acid reacts with an alcohol (or a compound containing an —OH group):

$$\underset{\text{Acid}}{R_1{-}C({=}O){-}OH} + \underset{\text{Alcohol}}{H{-}O{-}R_2} \longrightarrow \underset{\text{Ester}}{R_1{-}C({=}O){-}O{-}R_2} + H_2O \tag{31.1}$$

where R_1 and R_2 represent alkyl or aryl groups, such as CH_3—, C_2H_5—, or C_6H_5—.

High-molar-mass esters such as aspirin are generally insoluble in water and can be separated from a reaction mixture by crystallization. Aspirin can be prepared by the reaction of salicylic acid with acetic acid:

$$\underset{\text{Acetic acid}}{CH_3{-}C({=}O){-}OH} + \underset{\text{Salicylic acid}}{HO{-}C_6H_4{-}C({=}O){-}OH} \rightleftharpoons \underset{\text{Aspirin}}{CH_3{-}C({=}O){-}O{-}C_6H_4{-}C({=}O){-}OH} + H_2O \tag{31.2}$$

As the double arrow indicates, the reaction does not go to completion, but reaches equilibrium.

A better preparative method — the one you will use in this experiment — employs acetic anhydride instead of acetic acid. Acetic anhydride may be considered as the product of a reaction in which two acetic acid molecules combine, with the resulting elimination of a water molecule:

$$CH_3{-}C({=}O){-}OH + HO{-}C({=}O){-}CH_3 \longrightarrow CH_3{-}C({=}O){-}O{-}C({=}O){-}CH_3 + H_2O \tag{31.3}$$

The anhydride reacts with salicylic acid to yield the ester (aspirin):

$$CH_3-C(=O)-O-C(=O)-CH_3 + 2\ HO-C_6H_4-C(=O)OH \longrightarrow 2\ CH_3-C(=O)-O-C_6H_4-C(=O)OH + H_2O \qquad (31.4)$$

Acetic anhydride *(MM = 102)* | *Salicylic acid* *(MM = 138)* | *Aspirin* *(MM = 180)*

Excess anhydride reacts with the water produced in the esterification, thereby shifting the equilibrium in the forward direction and giving a better yield of the desired product. A catalyst, normally sulfuric or phosphoric acid, is used to increase the rate of the reaction.

SAFETY PRECAUTIONS AND DISPOSAL METHODS

Both acetic anhydride and phosphoric acid are reactive chemicals that can produce a serious burn on contact with the skin. In case of contact with either, wash the skin thoroughly with soap and water. Avoid breathing acetic anhydride vapors. Wash any spillage from the desk top. The aspirin you will prepare in this experiment is relatively impure and should not be taken internally.

Dispose of any excess solid chemical in a special container. Do not pour acetic anhydride down the drain. Follow the directions given by your instructor.

PROCEDURE

1. PREPARATION OF ASPIRIN

A) Preweigh a 50-mL Erlenmeyer flask on a decigram balance. Add 1.9 to 2.2 g of salicylic acid and weigh the flask again.

B) Pour 5.0 to 5.5 mL of acetic anhydride into the flask in such a way as to wash down any crystals of salicylic acid that may have adhered to the walls.

C) Add 5 drops of concentrated phosphoric acid (85 percent) to serve as a catalyst.

D) Clamp the flask in a beaker of water supported on a wire gauze (Figure 31–1). Heat the water to about 75°C, stirring the liquid in the flask occasionally with a stirring rod. Maintain this temperature for about 15 minutes, during which time the reaction should be complete.

E) *Cautiously* add 2 mL of water to the flask to decompose any excess acetic anhydride. Hot acetic acid vapor will evolve as a result of the decomposition.

F) When the liquid has stopped giving off vapors, remove the flask from the water bath and add 18 to 20 mL of water. Let the flask cool for a few minutes, during which time crystals of aspirin should begin to appear. Put the flask into an ice bath to hasten crystallization and increase the yield of the product. If crystals are slow to appear, it may be helpful to scratch the inside of the flask with a stirring rod.

G) Collect the aspirin by filtering the cold liquid through a Buchner funnel, using suction, as in Figure 31–2. Disconnect the rubber hose from the filter flask, pour about 5 mL of ice-cold deionized water over the crystals, and suck down the wash water. Repeat the washing step with a second 5 mL of ice-cold water. Draw air through the funnel for a few minutes to help dry the crystals, and then transfer them to a clean watch glass that has been preweighed to the nearest 0.1 g.

H) In order to determine the yield of aspirin in your experiment, it is necessary that the product be dry. If you do not have time to complete the experiment, store the watch glass carefully in your locker. At the beginning of the next laboratory period, weigh the watch glass and aspirin to the nearest 0.1 g. Record your data in the space provided in the report sheet.

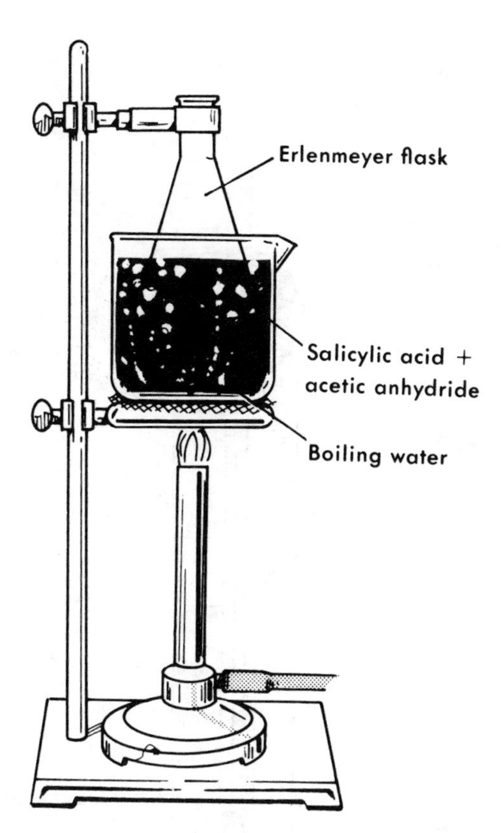

Figure 31–1.
Preparation of aspirin.

2. PURITY OF ASPIRIN (OPTIONAL)

Very pure aspirin melts at 135°C. By determining the melting point of your aspirin, you may estimate its purity, since the purer the aspirin, the closer its melting point will be to 135°C.

A) Assemble the apparatus shown in Figure 31–3, using a large oil-filled test tube as the heating bath.

B) Crush some of your aspirin crystals on a watch glass with a spatula. Form a mound from the powder and push the open end of a melting-point capillary into the mound. Hold the capillary vertically and allow it to drop against the table top, compacting the powder into a plug in the bottom of the tube. Repeating the process, build a plug about ¾ to 1 cm long.

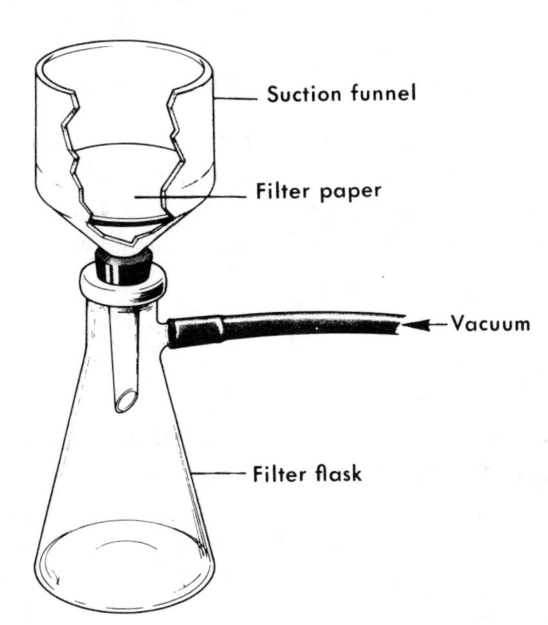

Figure 31–2. Vacuum filtration apparatus.

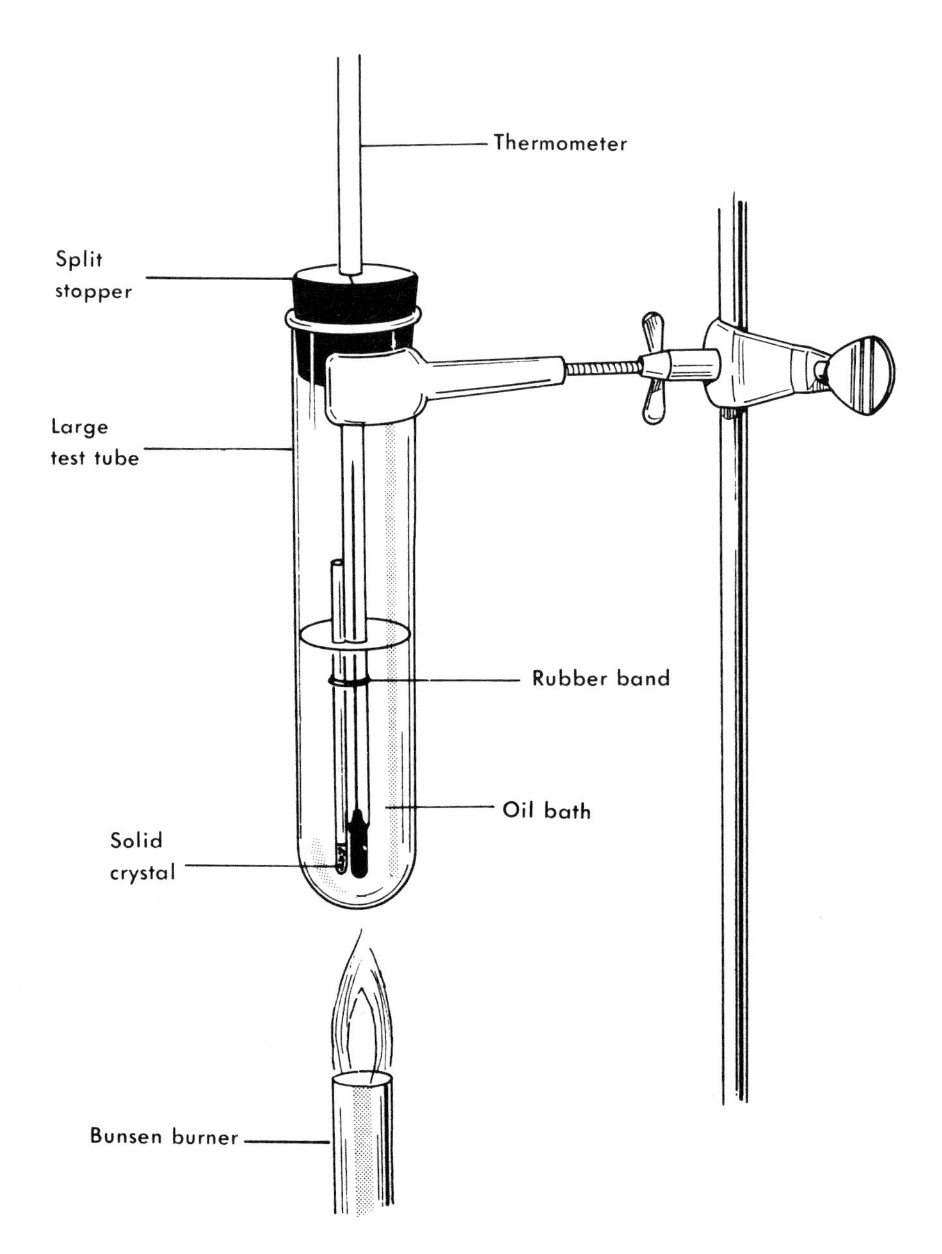

Figure 31–3. Apparatus for melting-point determination.

C) Attach the filled capillary to a thermometer with a rubber band or slice of rubber tubing, and immerse it in the oil bath. Do not allow the open end of the capillary to come into contact with the oil. Heat the bath rapidly with a Bunsen burner to about 100°C. As the melting point is approached, the crystals will begin to soften. Report the melting point as the temperature at which the last crystals disappear (the tube looks transparent).

CALCULATIONS

Based on the actual mass of salicylic acid used, calculate the theoretical yield of aspirin in grams, using Equation 31.4. Then determine the percentage yield,

$$\text{Percentage yield} = \frac{\text{actual yield}}{\text{theoretical yield}} \times 100$$

where actual yield means the number of grams of product actually obtained. Record your results on the report sheet.

Experiment 31
Work Page

NAME ....

DATE .... SECTION ....

1) PREPARATION OF ASPIRIN

A) Mass of 50-mL Erlenmeyer flask (g) ____________

B) Mass of flask and salicylic acid (g) ____________

C) Mass of salicylic acid (g) ____________

D) Mass of watch glass (g) ____________

E) Mass of watch glass and aspirin (g) ____________

F) Mass of aspirin (actual yield) (g) ____________

Theoretical yield of aspirin (show calculations below):

____________ g

Percentage yield (show calculations below):

____________ %

2) PURITY OF ASPIRIN (OPTIONAL)

Melting point of aspirin ____________°C

Experiment 31

Report Sheet

NAME ....

DATE .... SECTION ....

1) PREPARATION OF ASPIRIN

A) Mass of 50-mL Erlenmeyer flask (g) ____________

B) Mass of flask and salicylic acid (g) ____________

C) Mass of salicylic acid (g) ____________

D) Mass of watch glass (g) ____________

E) Mass of watch glass and aspirin (g) ____________

F) Mass of aspirin (actual yield) (g) ____________

Theoretical yield of aspirin (show calculations below):

____________ g

Percentage yield (show calculations below):

____________ %

2) PURITY OF ASPIRIN (OPTIONAL)

Melting point of aspirin ____________°C

Experiment 31

Advance Study Assignment

NAME ....

DATE .... SECTION ....

1) How would you test the purity of aspirin prepared in this experiment?

2) Calculate the theoretical yield of aspirin if you started with 1.75 g of salicylic acid.

3) Identify, by name or formula, R_1 and R_2 in Equation 31.1 when the ester *aspirin* is formed.

Experiment 32

Preparation and Properties of a Soap

Performance Goals

32–1 Starting with a vegetable oil, prepare a soap in the laboratory.

32–2 Examine the chemical properties of the soap you prepared.

CHEMICAL OVERVIEW

An ester is the product of the reaction between an alcohol and a carboxylic acid. The typical equation for the formation of an ester is

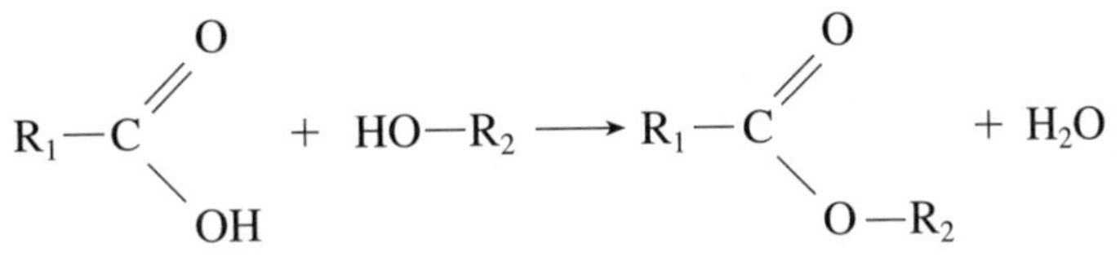

$$\underset{\text{Acid}}{R_1{-}C(=O){-}OH} + \underset{\text{Alcohol}}{HO{-}R_2} \longrightarrow \underset{\text{Ester}}{R_1{-}C(=O){-}O{-}R_2} + H_2O \qquad (32.1)$$

R_1 and R_2 are general symbols for *alkyl* groups containing only hydrogen and carbon. They may be the same group or they may be different. If the alcohol is glycerol, $C_3H_5(OH)_3$, and the acid is a long-chain fatty acid such as stearic acid, $C_{17}H_{35}COOH$, the ester is typical of those found in fats and oils. These esters can be reacted with strong bases to yield glycerol and the salt of the fatty acid. This process is known as *saponification,* and the sodium (or potassium) salts of the fatty acids are called *soaps*. The soap-making process may be written as

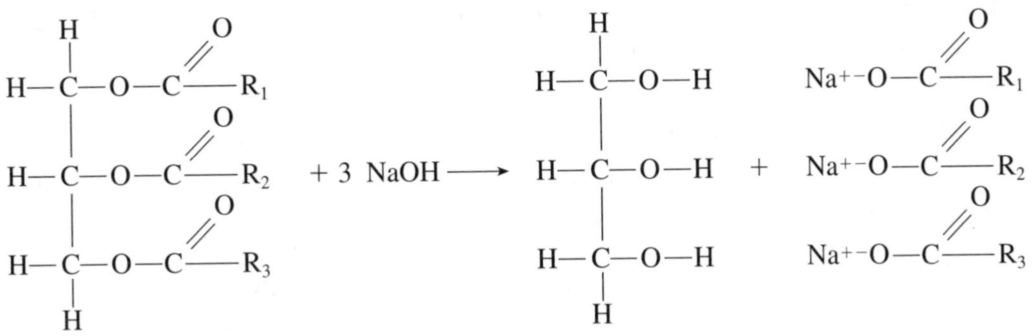

$$\begin{array}{l} H{-}C(H){-}O{-}C(=O){-}R_1 \\ H{-}C{-}O{-}C(=O){-}R_2 \\ H{-}C(H){-}O{-}C(=O){-}R_3 \end{array} + 3\ NaOH \longrightarrow \begin{array}{l} H{-}C(H){-}O{-}H \\ H{-}C{-}O{-}H \\ H{-}C(H){-}O{-}H \end{array} + \begin{array}{l} Na^{+-}O{-}C(=O){-}R_1 \\ Na^{+-}O{-}C(=O){-}R_2 \\ Na^{+-}O{-}C(=O){-}R_3 \end{array}$$

$$\textit{Ester} + \textit{Base} \longrightarrow \textit{Glycerol} + \textit{Soap} \qquad (32.2)$$

As you can see from Equation 32.2, the anions of soaps contain both polar groups ($—COO^-$) and nonpolar groups (long chains of carbon atoms). Polar compounds (or groups) are attracted to water and are called **hydrophilic.** Nonpolar compounds (or groups) are water-repelling or **hydrophobic,** and are soluble in or miscible with nonpolar compounds, such as fats, grease, oil, or other "dirt." This dual characteristic of soaps is the reason behind their cleaning action. The fat or oil is displaced from the fiber by the soap solution to form large globules that can be detached by jarring (rubbing) and then dispersed (emulsified) in the aqueous solution. **Emulsions** consist of fine droplets of one liquid dispersed in another liquid, in which the first is insoluble (like oil in water). The better the emulsion characteristics of a soap, the better it can clean.

Generally, soaps made from liquid fats (or oils) are more soluble than those made from solid fats. In the laboratory, you will prepare a soap by saponifying a vegetable oil with sodium hydroxide. Ethyl alcohol will be added to serve as a common solvent for the reactants. You will also investigate the characteristics of soaps formed from fatty acids and some divalent and trivalent cations. These cations are commonly encountered in areas where there is hard water and in industry. Their soaps are usually referred to as **metallic soaps.**

SAFETY PRECAUTIONS AND DISPOSAL METHODS

Alcohol vapors are highly flammable. Keep the flame away from the top of the beaker. Have an asbestos square handy to cover the beaker if the vapors should ignite. Also, sodium hydroxide causes severe burns when in contact with the skin. If a slippery feeling is observed on the hands or other parts of the body, rinse with large amounts of cold water immediately. Wear eye protection at all times during this experiment.

Dispose of excess vegetable oil and ethyl alcohol in separate bottles. Excess NaOH should be diluted with water and then poured down the drain. Collect solutions containing kerosene.

PROCEDURE

1. PREPARATION OF A SOAP

A) Weigh a 150-mL beaker on a decigram balance and weigh into it 18 to 20 g of vegetable oil. Add 20 mL of ethyl alcohol and 25 mL of 20% sodium hydroxide solution.

B) Stir the solution and support the beaker on an asbestos gauze on a tripod. Heat the beaker and its contents gently. Continue the heating until the odor of alcohol is no longer apparent and a pasty mass remains in the beaker. The reaction product is a mixture of the soap and the glycerol freed in the reaction (see Equation 32.2).

C) Allow the soap mixture to cool; then add 100 mL of saturated sodium chloride solution and stir thoroughly with a glass rod. This process is called "salting out" and is used to remove the soap from water, glycerol, and any excess sodium hydroxide present.

D) After the mixture has been stirred and mixed completely, filter off the soap on a Büchner funnel, using suction, as illustrated in Figure 31.2. Rinse with two 10-mL portions of ice-cold water, drawing this water through the funnel. Allow your soap to dry by spreading it out on a paper towel.

2. PROPERTIES AND REACTIONS OF SOAPS

A) **Washing properties.** Take a small amount of your soap and wash your hands with it. In soft water, it should lather easily. If any oil is left over, the soap will feel greasy. Describe the washing properties of your soap on the report sheet. Rinse your hands several times after the test.

B) **Basicity.** A soap that contains free alkali is harmful to the skin, silk, or wool. To test for the presence of free base, dissolve a small amount of your soap in 5 mL of ethyl alcohol and add two drops of phenolphthalein. If the indicator turns red, free alkali is present. Record your observation.

C) **Reaction with multivalent cations.** Dissolve about 1 g of your soap in 50 mL of warm water. Pour about 10 mL of soap solution into each of three test tubes. To the first test tube, add 8 or 10 drops of 5% $CaCl_2$; to the second, 8 or 10 drops of 5% $MgCl_2$; and to the third, 8 or 10 drops of 5% $FeCl_3$. Record your observations in the report sheet. (Does this remind you of the "scum" that forms when you wash in hard water?)

D) **Emulsification.** Put 5 to 10 drops of kerosene in a test tube containing 8 to 10 mL of water and shake it. An emulsion or suspension of tiny oil droplets in water will form (the solution will look cloudy). Let this solution stand for a few minutes. Prepare another test tube with the same ingredients, but add about 0.5 g of your soap to it before shaking it. Compare the stabilities of the emulsions in the two test tubes. Which emulsion seems to contain smaller droplets? Which emulsion clears up first? Explain. Record your answers on the report sheet.

Experiment 32

Work Page

NAME ....

DATE .... SECTION ....

A) WASHING PROPERTIES

Soap lathers a lot ______________ , a little ______________ , not at all ______________ .

Soap feels oily: yes ______________ , no______________ .

B) BASICITY

Soap solution + indicator: turns pink ______________ ; remains colorless ______________

C) REACTION WITH MULTIVALENT CATIONS

Cation Added *Observation*

Ca^{2+}

Mg^{2+}

Fe^{3+}

D) EMULSIFICATION

Emulsion containing smaller droplets:

Kerosene in water ______________ ; Kerosene + soap in water ______________ .

Emulsion that clears up first:

Kerosene in water ______________ ; Kerosene + soap in water ______________ .

Explanation:

Experiment 32

Report Sheet

NAME ....

DATE .... SECTION ....

A) WASHING PROPERTIES

Soap lathers a lot _____________ , a little _____________ , not at all _____________ .

Soap feels oily: yes _____________ , no_____________ .

B) BASICITY

Soap solution + indicator: turns pink _____________ ; remains colorless _____________

C) REACTION WITH MULTIVALENT CATIONS

Cation Added	*Observation*
Ca^{2+}	
Mg^{2+}	
Fe^{3+}	

D) EMULSIFICATION

Emulsion containing smaller droplets:

Kerosene in water _____________ ; Kerosene + soap in water _____________ .

Emulsion that clears up first:

Kerosene in water _____________ ; Kerosene + soap in water _____________ .

Explanation:

Experiment 32
Advance Study Assignment

NAME ....

DATE .... SECTION ....

1) What is a soap?

2) What does the term saponification mean?

3) What does the term emulsion mean?

4) What do the terms hydrophilic and hydrophobic mean?

Experiment 33

Carbohydrates

Performance Goals

33–1 Perform tests on various types of carbohydrates.
33–2 Identify an unknown carbohydrate.

CHEMICAL OVERVIEW

Carbohydrates are a class of organic compounds composed of carbon, hydrogen, and oxygen. Carbohydrates include polyhydroxyaldehydes (aldoses) and polyhydroxyketones (ketoses) or more complex molecules that can be broken down to yield these compounds. Many of our most common foods, such as rice, bread, potatoes, and fruits, are rich in carbohydrates.

The formulas of carbohydrate molecules can be represented by open-chain structures or by ring structures (Figure 33–1). All simple sugars in their open-chain form contain either an aldehyde group, —CHO, or a carbonyl group, $\rangle C{=}O$. Glucose contains six carbons and has a —CHO group in its open-chain form therefore it is classified as an **aldohexose.** Fructose is a **ketose** since it contains a ketone (carbonyl) group. Since fructose has six carbon atoms it is referred to as a **ketohexose.** The predominant forms of glucose and fructose in solution are the ring structures.

Further classification of carbohydrates is based on the number of simplest molecules **(monosaccharides)** present. **Disaccharides,** so-called "double sugars," are the result of the combination of two monosaccharides. Common examples of disaccharides are lactose (milk sugar), sucrose (table sugar), and maltose. **Polysaccharides,** such as starch and cellulose, are very-large-molar-mass polymers composed of many monosaccharide units.

Disaccharides and polysaccharides can be broken down to smaller units by hydrolysis (reaction with water). Sucrose and starch can be hydrolyzed under strongly acidic conditions or by using a biological catalyst called an enzyme.

There are many spot tests available that allow identification of major carbohydrates. In this experiment we will be using the following tests:

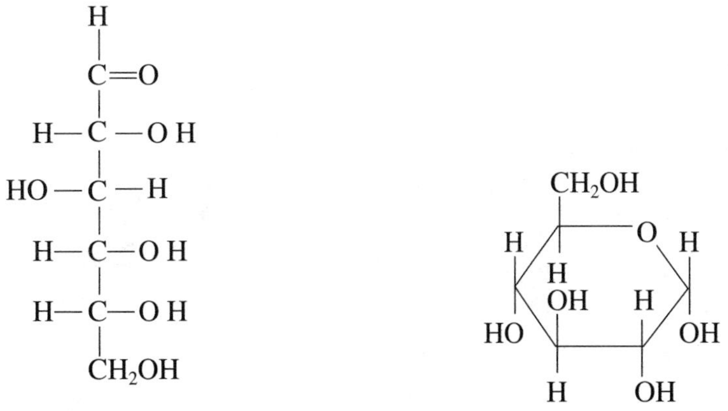

H
|
C=O
|
H—C—OH
|
HO—C—H
|
H—C—OH
|
H—C—OH
|
CH_2OH

D-glucose

CH_2OH
H O H
H
OH H
HO OH
H OH

α-D-glucopyranose

Figure 33–1. Open-chain and ring structures of glucose.

1. **Fehling's test.** This test is used to determine if sugars containing free aldehyde or ketone groups are present. These sugars, called **reducing sugars,** can react with mild oxidizing agents, such as Cu^{2+} in Fehling's solution, to yield Cu_2O, a red-orange solid.

$$Cu^{2+} + \text{reducing sugar} \rightarrow \underset{\text{(Red-orange)}}{Cu_2O(s)} + \text{oxidized sugar}$$

2. **Iodine test.** Polysaccharides, such as starch, produce a characteristic blue color when they react with a solution containing iodine. It is believed that a complex is formed that varies in intensity and shade of color depending on the size of the molecule, the concentration of iodine, and the temperature.
3. **Seliwanoff's test.** This test is used to differentiate between an aldohexose and a ketohexose. The *quick* appearance of a deep red color is a positive indication of a ketohexose, such as fructose. Aldohexoses react much more slowly.
4. **Barfoed's test.** The difference in reaction rate between reducing monosaccharides and reducing disaccharides with cupric acetate solution is the basis for this text. The smaller the sugar, the faster the reaction rate.

SAFETY PRECAUTIONS AND DISPOSAL METHODS

The reagents used in this experiment contain fairly concentrated acids, bases, and organic compounds that may cause burns or irritation to skin. Avoid contact with all solutions and wear eye protection while performing the experiment. Also, be careful when handling the boiling-water bath.

Solutions from this experiment can be poured down the drain unless otherwise directed by your instructor.

PROCEDURE

A boiling-water bath is required for some of the tests in this experiment. Pour about 200 mL of deionized water into a 400-mL beaker and heat it to boiling. Maintain it at that temperature, replenishing the water from time to time as it becomes necessary.

1. FEHLING'S TEST

Prepare the reagent for this test by mixing 20 mL of Fehling's solutions A and B. Add 1 mL of 1% solution of glucose, fructose, sucrose, lactose, and starch to separate, labeled test tubes, and then add 5 mL of the mixed Fehling's reagent to each test tube. Heat the test tubes in the boiling-water bath for 5 minutes and record your observations in the report sheet.

2. IODINE TEST

Add 1 mL of 1% solution of glucose, fructose, sucrose, lactose, and starch to separate, labeled test tubes, and then add 3 drops of dilute iodine solution to each. Prepare a sixth test tube with 1 mL of deionized water and 3 drops of iodine solution. Compare the colors observed and record them in the report sheet.

3. SELIWANOFF'S TEST

Dilute the fructose and glucose solutions available to 0.5% (2 mL sugar solution + 2 mL deionized water). Add 1 mL of each into separate test tubes, prepare a third one with 1 mL of deionized water. Add 10 mL of Seliwanoff's reagent to each test tube and place it into a boiling-water bath. Using a stopwatch (or second hand on your watch), record the time required for a color change to occur. Use the deionized water as a blank. Discontinue the heating after 10 minutes.

4. BARFOED'S TEST

Add 2 mL of 1% solution of fructose, glucose, and lactose to three separate, labeled test tubes. Add 5 mL of Barfoed's reagent to each test tube, mix, and place the test tubes into a boiling-water bath. Observe the time when a reaction becomes apparent (solution becomes cloudy or changes color). Discontinue heating after 15 minutes. Record your observations.

5. ANALYSIS OF AN UNKNOWN

Obtain an unknown solution from your instructor that may contain one of the following carbohydrates: fructose, glucose, sucrose, lactose, or starch. Using the tests you performed above, determine which compound is present in your solution. Remember, fructose and glucose are monosaccharides, sucrose is a nonreducing disaccharide, lactose is a reducing disaccharide, and starch is a polysaccharide.

Experiment 33
Work Page

NAME ....

DATE .... SECTION ....

1) FEHLING'S TEST

Substance	*Test Tube*	*Yes*	*No*
Glucose	1		
Fructose	2		
Sucrose	3		
Lactose	4		
Starch	5		
Unknown			

2) IODINE TEST

Substance	*Test Tube*	*Color*
Glucose	1	
Fructose	2	
Sucrose	3	
Lactose	4	
Starch	5	
H_2O	6	
Unknown		

3) SELIWANOFF'S TEST

Substance	*Test Tube*	*Yes*	*No*	*Time*
Fructose	1			
Glucose	2			
H_2O	3			
Unknown				

4) BARFOED'S TEST

Substance	*Test Tube*	*Time*
Fructose	1	
Glucose	2	
Lactose	3	
Unknown		

5) ANALYSIS OF AN UNKNOWN

Unknown Number ________________ Compound present ____________________________

Experiment 33
Report Sheet

NAME ....

DATE .... SECTION ....

1) FEHLING'S TEST

Substance	*Test Tube*	*Yes*	*No*
Glucose	1		
Fructose	2		
Sucrose	3		
Lactose	4		
Starch	5		
Unknown			

2) IODINE TEST

Substance	*Test Tube*	*Color*
Glucose	1	
Fructose	2	
Sucrose	3	
Lactose	4	
Starch	5	
H_2O	6	
Unknown		

3) SELIWANOFF'S TEST

Substance	*Test Tube*	*Yes*	*No*	*Time*
Fructose	1			
Glucose	2			
H_2O	3			
Unknown				

4) BARFOED'S TEST

Substance	*Test Tube*	*Time*
Fructose	1	
Glucose	2	
Lactose	3	
Unknown		

5) ANALYSIS OF AN UNKNOWN

Unknown Number ________________ Compound present ____________________________

Experiment 33

Advance Study Assignment

NAME ....

DATE .... SECTION ....

1) Complete the table below:

Substance	*Fehling's Test*		*Iodine Test*		*Seliw. Test*		*Barfoed's Test*	
	Yes	*No*	*Yes*	*No*	*Yes*	*No*	*Fast*	*Slow*
Glucose								
Fructose								
Sucrose					x	x		
Lactose					x	x		
Starch					x	x	x	x

2) What group of compounds gives a positive iodine test? Give an example.

3) (Optional) Sucrose is a disaccharide of glucose and fructose. Honey is a mixture of glucose and fructose. Explain why honey is a quicker energy source than table sugar (sucrose).

Experiment 34

Amino Acids and Proteins

Performance Goal

34–1 Perform identification tests on amino acids and proteins.

CHEMICAL OVERVIEW

Amino acids are molecules containing two functional groups: the amino group (—NH_2) and the carboxyl group (—COOH). Proteins are large, complex molecules, built from amino acids that are joined together by peptide linkages. A peptide bond is formed when an acidic carboxyl group of one amino acid reacts with the basic amino group of another amino acid molecule. The formation of such a bond is shown in Equation 34.1.

$$R_1{-}CH(NH_2){-}C({=}O){-}OH + H_2N{-}CH(R_2){-}C({=}O){-}OH \xrightarrow{-H_2O} R_1{-}CH(NH_2){-}\underbrace{C({=}O){-}NH}_{\text{Peptide bond}}{-}CH(R_2){-}C({=}O){-}OH \quad (34.1)$$

The product still contains a free amino group, which gives basic properties to the molecule, and a free carboxyl group, which gives it acidic properties.

When two amino acids react with each other the product is a **dipeptide.** When a very large number of amino acids are linked together, the **polypeptides** that form are known as **proteins.** These molecules make up our skin, muscles, and enzymes. Hormones, hair, and fingernails are also made of protein. In order for these molecules to function, they must exist in a specific, three-dimensional structure. Biological activity will cease if this structure is destroyed. Extreme temperatures, acids, bases, and heavy metal compounds can break linkages in proteins, causing them to become **denatured.**

In this experiment you will carry out some tests that are characteristic of amino acids and will investigate the various means of coagulating (denaturing) proteins.

1. **Biuret test.** This test identifies a compound that contains two or more peptide bonds. When these compounds react with Cu^{2+} ions in a basic solution, a pink-violet complex is formed (a positive test). Amino acids and dipeptides (two amino acids joined by one peptide bond) do not give violet colors with Cu^{2+}, but produce a blue solution (a negative test).
2. **Xanthoproteic test.** Some amino acids and the proteins containing them have aromatic rings. These rings react with concentrated nitric acid to produce a yellow compound, which is intensified in a basic solution. The test is used to determine the presence or absence of aromatic ring structures.
3. **Hopkins–Cole test.** This is a specific test for the presence of proteins containing triptophan, one of the essential amino acids. Eggs, for example, are high in triptophan. When a triptophan-containing protein solution is layered on top of concentrated sulfuric acid, a purple compound (ring) forms at the interface.

4. **Unoxidized sulfur test.** This test is to detect the presence of sulfur-containing amino acids (such as cysteine) or proteins containing these amino acids. In a basic solution inorganic sulfide ions, S^{2-}, are produced, which react with lead acetate to give PbS, a black precipitate.

SAFETY PRECAUTIONS AND DISPOSAL METHODS

The reagents used in this experiment contain organic compounds that may irritate skin. Concentrated nitric and sulfuric acids are *very corrosive.* Be sure not to get them on your skin or clothing. If you have spilled concentrated sulfuric acid on your skin, *wipe it off first,* then wash with soap and water. *WEAR EYE PROTECTION THROUGHOUT THE EXPERIMENT.* Also, handle the hot-water bath carefully.

Discard solutions containing heavy-metal ions in a stoppered bottle.

PROCEDURE

A boiling-water bath is required for some of the tests in this experiment. Pour about 200 mL of deionized water into a 400-mL beaker and heat it to boiling. Maintain it at that temperature, replenishing the water as it becomes necessary.

1. BIURET TEST

Label four clean test tubes. Pour 2 mL of egg albumin solution, 1% gelatin solution, 1% casein solution, and 0.5% alanine solution (an amino acid) into separate test tubes. Now add 3 mL of 10% sodium hydroxide solution to each one and shake carefully to mix. Add 2 drops of 2% copper sulfate solution to each test tube, mix, and observe the color of the solution. Record your observations in the report sheet.

2. XANTHOPROTEIC TEST

Pour 2 mL of egg albumin solution into a test tube and add 10 drops of concentrated nitric acid to it. Mix and heat in a boiling-water bath for 2 minutes. Record any color change. Cool the mixture and add 10% sodium hydroxide solution dropwise until the mixture is basic to litmus (red litmus turns blue). Note any change in color. Record your observations in the report sheet.

3. HOPKINS–COLE TEST

Label three clean test tubes. Pour 2 mL of egg albumin solution, 1% triptophan solution, and 2% gelatin solution into separate test tubes. Add to each one 2 mL of Hopkins–Cole reagent and mix. Now, holding the first test tube at a 45° angle, *VERY CAREFULLY AND SLOWLY,* add 30 drops of concentrated sulfuric acid down the inside wall of the test tube, so that it forms a layer on the bottom. Avoid shaking the tube or doing anything that will cause mixing. The presence of a purple ring at the interface is a positive test for triptophan. If no ring forms, gently tap the test tube to cause slight mixing at the interface. Record your observations on the report sheet. Repeat the procedure for the second and third test tubes.

4. UNOXIDIZED SULFUR TEST

Pour 2 mL of egg albumin solution into a test tube. Add 5 mL of 10% sodium hydroxide solution and 3 drops of 5% lead acetate solution. Mix and heat in a boiling-water bath for 3 minutes. Observe the results and record them in the report sheet.

5. COAGULATION/PRECIPITATION OF PROTEINS

Obtain five test tubes and pour 3 mL of egg albumin solution into each one.

A) Heat the first test tube to boiling. Observe any changes and record your results in the report sheet.
B) To the second test tube add 10 mL of 95% ethanol (ethyl alcohol). Mix and observe changes. Record your observations in the report sheet.
C) To the third test tube add 2% silver nitrate solution dropwise. Record your observations.
D) Repeat the preceding procedure using the fourth sample and 5% mercury(II) chloride solution.
E) Repeat the procedure on the fifth sample, using dilute tannic acid.

Experiment 34
Work Page

NAME ....

DATE .... SECTION ....

1) BIURET TEST

Test Tube	*Substance*	*Yes*	*No*
1	Egg albumin		
2	Gelatin		
3	Casein		
4	Alanine		

If you heated a protein solution with an acid to break the bonds and to produce the amino acid "building blocks," would you expect the hydrolysis product to give a positive ___________ or negative ___________ Biuret test? Explain.

2) XANTHOPROTEIC TEST

Original color of solution: ___________________________

Color after heating with HNO_3: ___________________________

Color of basic solution: ___________________________

3) HOPKINS–COLE TEST

Test Tube	*Substance*	*Observation*
1	Egg albumin	
2	Triptophan	
3	Gelatin	

4) UNOXIDIZED SULFUR TEST

Formation of a precipitate: Yes _________ ; No _________

Write the equation for the reaction, if any.

5) COAGULATION/PRECIPITATION OF PROTEINS

Part	*Observation*
A	
B	
C	
D	
E	

6) QUESTIONS (OPTIONAL)

A) What happens to egg albumin when an egg is hard-boiled?

B) A dilute solution of mercury(II) chloride can be used to preserve anatomical specimens. Explain why.

C) What happens chemically to proteins in your body during digestion?

Experiment 34
Report Sheet

NAME ....

DATE .... SECTION ....

1) BIURET TEST

Test Tube	*Substance*	*Yes*	*No*
1	Egg albumin		
2	Gelatin		
3	Casein		
4	Alanine		

If you heated a protein solution with an acid to break the bonds and to produce the amino acid "building blocks," would you expect the hydrolysis product to give a positive __________ or negative __________ Biuret test? Explain.

2) XANTHOPROTEIC TEST

Original color of solution: ____________________

Color after heating with HNO_3: ____________________

Color of basic solution: ____________________

3) HOPKINS–COLE TEST

Test Tube	*Substance*	*Observation*
1	Egg albumin	
2	Triptophan	
3	Gelatin	

4) UNOXIDIZED SULFUR TEST

Formation of a precipitate: Yes _________ ; No _________

Write the equation for the reaction, if any.

5) COAGULATION/PRECIPITATION OF PROTEINS

Part	*Observation*
A	
B	
C	
D	
E	

6) QUESTIONS (OPTIONAL)

A) What happens to egg albumin when an egg is hard-boiled?

B) A dilute solution of mercury(II) chloride can be used to preserve anatomical specimens. Explain why.

C) What happens chemically to proteins in your body during digestion?

Experiment 34

Advance Study Assignment

NAME ....

DATE .... SECTION ....

1) Which two functional groups are present in all amino acids?

2) Explain why your skin turns yellow if it comes in contact with concentrated nitric acid.

3) What process takes place when milk goes sour?

4) (Optional) Why is alcohol used as a disinfectant before you get a shot?

5) (Optional) When a person is suspected to have ingested a heavy-metal salt, such as $AgNO_3$ or $Pb(NO_3)_2$, why is egg white or milk used as an antidote?

Appendix

The Oxidation Numbers of Some Common Cations

Ionic Charge: +1		***Ionic Charge: +2***		***Ionic Charge: +3***	
Alkali Metals: Group IA		*Alkaline Earths: Group IIA*		*Group IIIA*	
Li^{+}	Lithium	Be^{2+}	Beryllium	Al^{3+}	Aluminum
Na^{+}	Sodium	Mg^{2+}	Magnesium	Ga^{3+}	Gallium
K^{+}	Potassium	Ca^{2+}	Calcium	*Transition Elements*	
Rb^{+}	Rubidium	Sr^{2+}	Strontium	Cr^{3+}	Chromium(III)
Cs^{+}	Cesium	Ba^{2+}	Barium	Mn^{3+}	Manganese(III)
Transition Elements		*Transition Elements*		Fe^{3+}	Iron(III)
Cu^{+}	Copper(I)	Cr^{2+}	Chromium(II)	Co^{3+}	Cobalt(III)
Ag^{+}	Silver	Mn^{2+}	Manganese(II)		
Polyatomic Ions		Fe^{2+}	Iron(II)		
NH_4^{+}	Ammonium	Co^{2+}	Cobalt(II)		
		Ni^{2+}	Nickel		
Others		Cu^{2+}	Copper(II)		
H^{+}	Hydrogen	Zn^{2+}	Zinc		
or		Cd^{2+}	Cadmium		
H_3O^{+}	Hydronium	Hg_2^{2+}	Mercury(I)		
		Hg^{2+}	Mercury(II)		
		Others			
		Sn^{2+}	Tin(II)		
		Pb^{2+}	Lead(II)		

The Oxidation Numbers of Some Common Anions

Ionic Charge: −1				*Ionic Charge: −2*		*Ionic Charge: −3*	
Halogens: Group VIIA		*Oxyanions*		*Group VIA*		*Group VA*	
F^-	Fluoride	ClO_4^-	Perchlorate	O^{2-}	Oxide	N^{3-}	Nitride
Cl^-	Chloride	ClO_3^-	Chlorate	S^{2-}	Sulfide	P^{3-}	Phosphide
Br^-	Bromide	ClO_2^-	Chlorite	*Oxyanions*		*Oxyanion*	
I^-	Iodide	ClO^-	Hypochlorite	CO_3^{2-}	Carbonate	PO_4^{3-}	Phosphate
Acidic Anions				SO_4^{2-}	Sulfate		
HCO_3^-	Hydrogen carbonate	BrO_3^-	Bromate	SO_3^{2-}	Sulfite		
HS^-	Hydrogen sulfide	BrO_2^-	Bromite	$C_2O_4^{2-}$	Oxalate		
HSO_4^-	Hydrogen sulfate	BrO^-	Hypobromite	CrO_4^{2-}	Chromate		
HSO_3^-	Hydrogen sulfite	IO_4^-	Periodate	$Cr_2O_7^{2-}$	Dichromate		
$H_2PO_4^-$	Dihydrogen phosphate	IO_3^-	Iodate	*Acidic Anion*			
Other Anions		NO_3^-	Nitrate	HPO_4^{2-}	Monohydrogen phosphate		
SCN^-	Thiocyanate	NO_2^-	Nitrite	*Diatomic*			
CN^-	Cyanide	OH^-	Hydroxide	O_2^{2-}	Peroxide		
H^-	Hydride	$C_2H_3O_2^-$	Acetate				
		MnO_4^-	Permanganate				

Concentration of Desk Reagents

Reagent	*Formula*	*Molarity*	*% Solute*
Hydrochloric acid, conc.	HCl	12 M	37
Hydrochloric acid, dil.		6	20
Nitric acid, conc.	HNO_3	16	71
Nitric acid, dil.		6	32
Sulfuric acid, conc.	H_2SO_4	18	96
Sulfuric acid, dil.		3	25
Acetic acid, glacial	$HC_2H_3O_2$	17	99.5
Acetic acid, dil.		6	34
Aqueous ammonia, conc.	$NH_3(aq)$	15	29
Aqueous ammonia, dil.		6	12
Sodium hydroxide, dil.	$NaOH$	6	20

DATE **EXPERIMENT** **NAME**

DATE **EXPERIMENT** **NAME**